Singlet Oxygen Detection and Imaging

Synthesis Lectures on Materials and Optics

Singlet Oxygen Detection and Imaging

Steffen Hackbarth, Michael Pfitzner, Jakob Pohl, and Beate Röder

ISBN: 978-3-031-01263-1   paperback
ISBN: 978-3-031-02391-0   ebook
ISBN: 978-3-031-00255-7   hardcover

DOI: 10.1007/978-3-031-02391-0

A Publication in the Springer series
SYNTHESIS LECTURES ON MATERIALS AND OPTICS

Lecture #5
Series ISSN
Synthesis Lectures on Materials and Optics
Print 2691-1930   Electronic 2691-1949

Singlet Oxygen Detection and Imaging

Steffen Hackbarth, Michael Pfitzner, Jakob Pohl, and Beate Röder
Humboldt-Universität zu Berlin

SYNTHESIS LECTURES ON MATERIALS AND OPTICS #5

ABSTRACT

Singlet Oxygen, the lowest electronically excited state of molecular oxygen, is highly reactive and involved in many chemical and biological processes. It is one major mediator during photosensitization, which has been used by mankind since ancient times, even though the mechanisms behind it were understood only about half a century ago.

The combination of high reactivity and very long natural lifetime allows for direct optical detection of singlet oxygen and its interactions using its characteristic phosphorescence at around 1270 nm. Since this emission is very weak, optical detection was technically very challenging for a long time. Therefore, even today, most laboratories only exploit the high reactivity to observe the interaction with sensor molecules, rather than singlet oxygen emission itself. However, in recent years highly sensitive optical detection was developed, the authors being major contributors.

This book is dedicated to the detection of singlet oxygen, discussing possibilities, pitfalls and limits of the various methods with a special focus on time-resolved phosphorescence and the kinetics of singlet oxygen generation and decay including involved and related processes, discussing investigated systems with various complexity from solutions over *in vitro* to *in vivo*.

The long-standing paradigm that singlet oxygen phosphorescence is a benchmark for detection systems rather than an option for process observation is still ubiquitous and this book hopes to contribute in overcoming this still prevailing bias.

KEYWORDS

singlet oxygen, NIR, photosensitizer, time-resolved spectroscopy, photodynamic inactivation, photodynamic therapy, in vivo, in vitro, CAM, bacteria

Contents

PART III $^1O_2$ in Biological Systems 45

PART I

General

CHAPTER 1

Introduction

In this book, we will discuss the historic developments and current state of the art concerning the detection of singlet oxygen ($^1O_2$) generated through photosensitization. In addition to various detection methods including their technical requirements and limitations, this book contains detailed considerations on the kinetics of $^1O_2$ in solution and systems of higher complexity as well as an overview of current achievements in the field of time-resolved detection of $^1O_2$ in biological systems, from cell suspensions over biofilms up to *in vivo* systems.

1.1 SIGNIFICANCE AND SOCIAL IMPORTANCE

Humanity has been using photosensitization for thousands of years without understanding the underlying mechanism. Rubbing with certain herbs in combination with sunlight can treat skin impurities, as was already known in ancient Egypt. After observation of many similar interactions between light and matter, the concept of photosensitization was developed: a drug-mediated reaction of biological material to light. The drugs involved in such processes were later referred to as photosensitizers (PSs).

The first documented scientific studies began in 1897 when Oscar Raab analyzed the toxicity of several fluorescent dyes to infusoria for the treatment of malaria. Together with Herrmann von Tappheimer, he discovered that acridine orange in combination with light can kill plasmids and first indications of a connection of this "photodynamic action" with oxygen were also found [1]. Both scientists recognized the therapeutic potential of this discovery already in 1903 [2]. They published first suggestions for the use of the photodynamic effect for the treatment of skin diseases. In 1905, Albert Jesionek documented the first photodynamic therapy (PDT) of a patient, treating skin lesions of a 70-year-old woman with eosin and sunlight [3].

Some years later, Meyer-Betz carried out his famous self-experiment during his investigations of porphyrin metabolism diseases, demonstrating that there is a direct relation between skin photosensitization and an increased amount of free porphyrins in the blood stream [4]. These two findings were milestones on the way to understanding and applying the photodynamic effect to biological systems. In principle, they opened up a new path for the application of tetrapyrroles and other dyes in a gentle, noninvasive therapy of different diseases. These first reports were followed by broad research activity in the 1940s and 1950s with a lot of interesting results [5, 6].

Nevertheless, the use of the photodynamic effect was limited to skin diseases at this time. In the mid-1970s, the photodynamic effect became of greater interest for medicine due to the

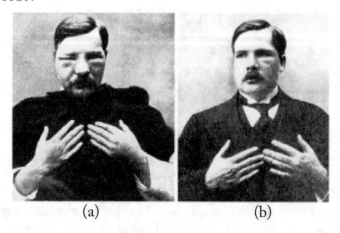

(a)   (b)

Figure 1.1: Meyer-Betz during and after his self-experiment with 200 mg hematoporphyrin and sunlight [4].

rapid development of lasers and fiber optics. With these new techniques, it was now possible to treat all surfaces in the human body attainable through fiber optics with a diameter of some hundreds of micrometers. This included the development of noninvasive methods for the treatment of cancer [7], skin diseases like psoriasis [8], or its use for the photoinduced induction of DNA strain breaks [9].

In addition to its use in PDT and photodynamic inhibition (PDI), $^1O_2$ is a very versatile actor as it has influence on many important processes. Reactive oxygen species like $^1O_2$ mediate signal transduction processes for the expression of several proteins [10], gene regulation via the transcription factor activator protein-2 [11], and activate protein kinases [12]. They also play a role during photosynthesis in the photosystem of bacteria, algae, and higher plants. Furthermore, $^1O_2$ is regarded to play the key role in both the apoptotic and necrotic pathways of cell death induced by PDT, as well as in the inhibition of microorganisms of all kinds (PDI). Moreover, it is also an important actor during chromophore degradation, e.g., in solar cells [13], in OLEDs [14] as well as in skin-aging effects [15] and $^1O_2$ generated by humic substances in surface waters, soils, and swamps has an influence on aquatic organisms [16] (Section 6.3).

1.2   MOLECULAR MECHANISM OF PHOTODYNAMIC THERAPY

The principle of $^1O_2$ generation was first proposed by Kautsky in numerous publications from 1931–1939 [17]. Kautsky was the first to discover the interaction between phosphorescent dyes and oxygen as mediators of autooxidation or photooxidation, as it was called then. In 1939, he proposed $^1O_2$ to be the intermediate species in dye-sensitized photooxidation [18]. Many noted scientists of that time rejected the idea of $^1O_2$ molecules as mobile, reactive intermediates. Evidence for the energy transfer from a PS in triplet state to molecular oxygen was first given by Snelling [19]. In the same year, Christopher Foote gave a schematic summary of the subsequent discussions and investigations concerning the generation mechanism of $^1O_2$ [20].

He postulated two types of photosensitization, both occurring in the presence of molecular oxygen. The "Type I" process includes charge transfer while the "Type II" path based on the photosensitized formation of $^1O_2$ involves energy transfer processes. Later, Laustriat postulated a third way of photosensitization, "Type III", which occurs in the absence of molecular oxygen and involves charge transfer [9]. In view of this broad definition of photosensitization, photoinduced electron transfer processes (e.g., photosynthesis) can also be treated as photosensitized reactions.

The processes of Types I and II photosensitization take place simultaneously in a system, which means that they compete with each other. The relative efficiency of both pathways is mainly defined by the physicochemical properties of the system. At normal oxygen concentration and neutral to physiological pH, the Type II mechanism will dominate in most cases. Using model systems, a distinction between the two types of photosensitization can be made on the base of different photoproducts.

1.3   PHOTOSENSITIZED GENERATION OF SINGLET MOLECULAR OXYGEN

The special properties and the important role of molecular oxygen results from its extraordinary electronic structure. Since this molecule has two unpaired electrons, the ground state of molecular oxygen is a triplet state (T_0)—in contrast to the majority of molecules, which have a "closed shell" conformation in the ground state (S_0). The first excited electronic state of molecular oxygen, however, is a singlet state (S_1: $^1\Delta_g$ so-called: $^1O_2$). This means that $^1O_2$ is generated as a result of a spin-forbidden inter system crossing (ISC) process, making direct excitation of this state with laser irradiation possible but at very low quantum yields [21].

Only a very small amount of energy (0.98 eV) is needed for the generation of $^1O_2$ and its lifetime strongly depends on the environment. In solution, it can adopt values between 700 µs in CCl_4 and 3.6 µs in aqueous solutions. To excite the second excited singlet state ($^1\Sigma_g^+$), which has a very short lifetime (about 0.1 ns in solution), 1.6 eV is needed. The first excited triplet state ($^3\Sigma_g^+$) is located in the UV region. Because of these energetic conditions, only the first excited singlet state $^1\Delta_g$ of molecular oxygen ($^1O_2$) is of importance for biomedical and environmental applications.

The processes relevant for photosensitization occurring after light absorption by the PS molecule are summarized in Figure 1.2. After absorption of a light quantum, the PS is excited directly to the first excited singlet state (S_1) or passes from a higher excited singlet state (S_n) to the S_1-state. In general, energy and electron transfer processes can occur from this excited electronic state. Nevertheless, the most important step in photosensitization is the ISC between

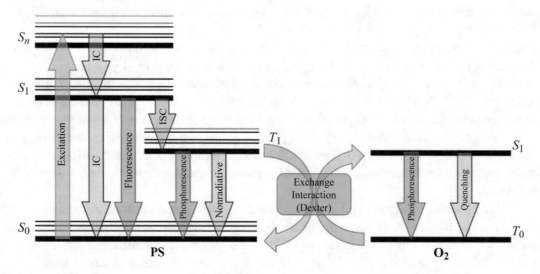

Figure 1.2: Jablonski diagram, showing transitions relevant for $^1O_2$ generation.

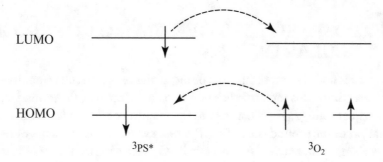

Figure 1.3: Dexter transfer.

the S_1 state of the PS to its first excited triplet state (T_1). The triplet state is quite long lived (up to some hundred microseconds), making diffusion controlled interactions like energy transfer to molecular oxygen possible. The energy transfer from a PS in triplet state to molecular oxygen (T_0), both changing to singlet state, cannot happen by dipole-dipole interaction. The interaction is a Dexter electron transfer, during which the PS hands over an excited LUMO electron to the oxygen. Oxygen returns an electron to the PS HOMO (Figure 1.3). The transfer strongly depends on the distance between the involved molecules, making the whole process diffusion limited. Since the electronic wave functions need to overlap, close contact is required. As electrons are indistinguishable, this exchange is an energy transfer. In principle, however, we discuss an electron exchange without spin flip, so that the process is not subject to any limitations apart from more or less matching energy levels and spatial proximity.

A useful tool for estimating the potential photosensitizing efficiency of a given compound is the determination of its singlet oxygen quantum yield (Φ_Δ). This value can be obtained from a kinetic analysis of the relevant processes shown in Figure 1.2. As a result, we get:

$$\Phi_\Delta = \frac{k_{ISC}}{k_{ISC} + k_{IC} + k_{Flu}} \cdot \frac{k_{EnT}[O_2]}{k_{EnT}[O_2] + k_{Phos} + k_{NR}} \cdot S_\Delta \tag{1.1}$$

with the rate constants (k) of the ISC, internal conversion (IC), radiative deactivation (Flu or $Phos$), non-radiative deactivation of T_1 which strictly spoken is also an ISC (NR), and energy transfer (EnT). $[O_2]$ is the concentration of molecular oxygen. The factor S_Δ takes into account that $^1O_2$ is not the only possible result of an interaction between the PS in its triplet state and molecular oxygen. For most molecules, the value of S_Δ is much smaller than one but for tetrapyrroles it approaches nearly one [22]. Simplifying Equation (1.1) we finally get:

$$\Phi_\Delta = \Phi_{ISC} \cdot \tau_T \cdot k_{EnT}[O_2] \cdot S_\Delta \tag{1.2}$$

From this equation it can be deduced that Φ_Δ depends on the triplet quantum yield (Φ_{ISC}) as well as the triplet lifetime (τ_T) of the PS under the given experimental conditions. This lifetime strongly dependens on the oxygen concentration and other environmental factors.

The transition of $^1O_2$ to the ground state is three-fold forbidden according to different selection rules. This results in a very low rate constant of the radiative deactivation and the observed lifetime mainly depends on nonradiative deactivation processes. Depending on the microenvironment, the phosphorescence quantum yields range between 10^{-5} and 10^{-6} [23].

While reactions of molecular oxygen with singlet state organic molecules are spin-forbidden, $^1O_2$ readily reacts with unsaturated organic molecules in a spin-allowed process [24]. Therefore, many biological components—first of all proteins—react with $^1O_2$ [25], which finally results in the cell-toxicity of PDT.

Some $^1O_2$ quenchers, like e.g., 1,3-diphenylisobenzofuran (DPBF), reach remarkable reaction rates and the quencher consumption or appearance of reaction products can act as a measure for the amount of generated $^1O_2$ [26]. However, since every chemical reaction competes with other deactivation processes, the reaction rate depends strongly on the setup. Nevertheless, chemical quenching can be a reliable method to determine the amount of generated $^1O_2$ in well-chosen systems like homogeneous organic solutions. Furthermore, the substance under observation can be quantified, using standard laboratory equipment, by absorption [26], fluorescence [27], or ESR [28]. Unfortunately, in heterogeneous systems this method soon reaches its limits in respect to quantitative accuracy [29]. Furthermore, time-resolved $^1O_2$ measurements are not possible this way since the speed of the aforementioned detection methods is insufficient.

The first spectroscopic detection of $^1O_2$ was accomplished by Khan and Kasha [32]. They observed the oxygen dimol emission—an intensive red glow at 633 and 703 nm—following the decomposition of hydrogen peroxide in the presence of sodium hypochlorite. Less than a year later, Bader and Ogryzlo published the detection of a weak signal at 1270 nm which was addressed to the (0,0) transition $^1\Delta_g^+ \rightarrow ^3\Sigma_g^-$ [33].

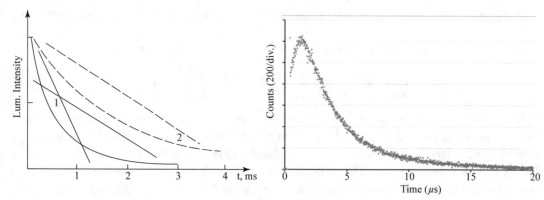

Figure 1.4: $^1O_2$ kinetics as they were recorded in 1979 in CCl_4 with flash lamp excitation [30] in comparison to a measurement of 5,10,15,20-Tetrakis(3-hydroxyphenyl)chlorin (mTHPC) in Jurkat cells performed in 2014 using LED excitation at 402 nm (20 s, 2mW) [31].

Just a benchmark for the sensitivity of spectroscopic instruments for a long time, today the time-resolved detection of this extremely weak signal has become a versatile tool for the investigation of photosensitized effects in materials and complex biological systems. The omnipresence of oxygen, the defined generation of $^1O_2$ by photosensitizers and its high reactivity, makes it a universal sensor, limited only by the sensitivity of the available detection system. However, sensitivity has grown by orders of magnitude during the last decades (see Figure 1.4) and there is still room for improvement.

REFERENCES

[1]   O. Raab. Über die Wirkung fluoreszierender Stoffe auf Infusorien. *Zeitschrift für Biologie* 39 (1900), pp. 524–526.

[2]   H. von Tappeiner. Therapeutische Versuche mit fluoreszierenden Stoffen. *Münchener Medizinische Wochenschrift* 1 (1903), pp. 2042–2044.

[3]   A. Jesionek and H. von Tappeiner. Zur Behandlung der Hautcarcinome mit fluoreszierenden Stoffen. *Deutsches Archiv für Klinische Medizin* 85 (1905), pp. 223–239.

[4]   F. Meyer-Betz. Untersuchungen über die biologische (photodynamische) Wirkung des Hämatoporphyrins und anderer Derivate des Blut- und Gallenfarbstoffes. *Deutsches Archiv für Klinische Medizin* 112 (1913), pp. 476–503.

[5]   H. Vogel. *Das Chlorophyll: in Medizin und Kosmetik*. Nürnberg: Hans Carl, 1954.

[6]   J. Brugsch. *Porphyrine*. Leipzig: F. A. Barth, 1959.

[7]   M. A. Trelles. *Laser tumour therapy: Second Plenary Workshop*. Cambrils and Madrid: Instituto Médico Vilafortuny and Ilustre Colegio Oficial de Médicos de la Autonomía de Madrid, 1990.

[8]   B. Röder et al. Mittel zur Behandlung von Hauterkrankungen und Tumoren. WP 248 282. 1984.

[9]   G. Laustriat. Molecular mechanisms of photosensitization. *Biochimie* 68.6 (1986), pp. 771–778. DOI: 10.1016/S0300-9084(86)80092-X.

[10]   A. Mahns et al. Contribution of UVB and UVA to UV-dependent stimulation of cyclo-oxygenase-2 expression in artificial epidermis. *Photochemical & Photobiological Sciences: Official Journal of the European Photochemistry Association and the European Society for Photobiology* 3.3 (2004), pp. 257–262. DOI: 10.1039/B309067A.

[11]   S. Grether-Beck et al. Activation of transcription factor AP-2 mediates UVA radiation- and singlet oxygen-induced expression of the human intercellular adhesion molecule 1 gene. *Proceedings of the National Academy of Sciences of the United States of America* 93.25 (1996), pp. 14586–14591. DOI: 10.1073/pnas.93.25.14586.

[12]   G. Kick et al. Strong and prolonged induction of c-jun and c-fos proto-oncogenes by photodynamic therapy. *British Journal of Cancer* 74.1 (1996), pp. 30–36. DOI: 10.1038/bjc.1996.311.

[13]   M. Jørgensen, K. Norrman, and F. C. Krebs. Stability/degradation of polymer solar cells. *Solar Energy Materials and Solar Cells* 92.7 (2008), pp. 686–714. DOI: 10.1016/j.solmat.2008.01.005.

[14]   S. Schmidbauer, A. Hohenleutner, and B. König. Studies on the photodegradation of red, green and blue phosphorescent OLED emitters. *Beilstein Journal of Organic Chemistry* 9 (2013), pp. 2088–2096. DOI: 10.3762/bjoc.9.245.

[15]   M. D. Carbonare and M. A. Pathak. Skin photosensitizing agents and the role of reactive oxygen species in photoaging. *Journal of Photochemistry and Photobiology B: Biology* 14.1-2 (1992), pp. 105–124. DOI: 10.1016/1011-1344(92)85086-A.

[16]   A. Paul et al. Photogeneration of singlet oxygen by humic substances: Comparison of humic substances of aquatic and terrestrial origin. *Photochemical & Photobiological Sciences: Official Journal of the European Photochemistry Association and the European Society for Photobiology* 3.3 (2004), pp. 273–280. DOI: 10.1039/B312146A.

[17]   H. Kautsky. Energie-Umwandlungen an Grenzflächen, IV. Mitteil: H. Kautsky und A. Hirsch: Wechselwirkung zwischen angeregten Farbstoff-Molekülen und Sauerstoff. *Berichte der deutschen chemischen Gesellschaft (A and B Series)* 64.10 (1931), pp. 2677–2683. DOI: 10.1002/cber.19310641017.

[18]   H. Kautsky. Quenching of luminescence by oxygen. *Transactions of the Faraday Society* 35 (1939), p. 216. DOI: 10.1039/TF9393500216.

[19]   D. R. Snelling. Production of singlet oxygen in the benzene oxygen photochemical system. *Chemical Physics Letters* 2.5 (1968), pp. 346–348. DOI: 10.1016/0009-2614(68) 80093-4.

[20]   C. S. Foote. Mechanisms of photosensitized oxidation. *Science* 162.3857 (1968), pp. 963–970. DOI: 10.1126/science.162.3857.963.

[21]   S. Jockusch et al. Singlet molecular oxygen by direct excitation. *Photochemical & Photobiological Sciences: Official Journal of the European Photochemistry Association and the European Society for Photobiology* 7.2 (2008), pp. 235–239. DOI: 10.1039/B714286B.

[22]   R. V. Bensasson, E. J. Land, and T. G. Truscott. *Excited States and Free Radicals in Biology and Medicine: Contributions from Flash Photolysis and Pulse Radiolysis.* Oxford University Press, 1993.

[23]   A. Gollmer et al. Singlet Oxygen Sensor Green (R): photochemical behavior in solution and in a mammalian cell. *Photochemistry and Photobiology* 87.3 (2011), pp. 671–679. DOI: 10.1111/j.1751-1097.2011.00900.x.

[24]   E. L. Clennan. Chapter 18. Overview of the Chemical Reactions of Singlet Oxygen. *Singlet Oxygen*, ed. by S. Nonell and C. Flors. Vol. 1. Comprehensive Series in Photochemical & Photobiological Sciences. Cambridge: Royal Society of Chemistry, 2016, pp. 351–367. DOI: 10.1039/9781782622208-00351.

[25]   M. J. Davies. Reactive species formed on proteins exposed to singlet oxygen. *Photochemical & Photobiological Sciences: Official Journal of the European Photochemistry Association and the European Society for Photobiology* 3.1 (2004), pp. 17–25. DOI: 10.1039/b307576c.

[26]   W. Spiller et al. Singlet oxygen quantum yields of different photosensitizers in polar solvents and micellar solutions. *Journal of Porphyrins and Phthalocyanines* 2.2 (1998), pp. 145–158.

[27]   C. Flors et al. Imaging the production of singlet oxygen in vivo using a new fluorescent sensor, Singlet Oxygen Sensor Green. *Journal of Experimental Botany* 57.8 (2006), pp. 1725–1734. DOI: 10.1093/jxb/erj181.

[28]   S. K. Han et al. Evidence of singlet oxygen and hydroxyl radical formation in aqueous goethite suspension using spin-trapping electron paramagnetic resonance (EPR). *Chemosphere* 84.8 (2011), pp. 1095–1101. DOI: 10.1016/j.chemosphere.2011.04.051.

[29]   S. Hackbarth, T. Bornhütter, and B. Röder. Singlet oxygen in heterogeneous systems. *Singlet Oxygen*, ed. by S. Nonell and C. Flors. Vol. 2. Comprehensive Series in Photochemical & Photobiological Sciences. The Royal Society of Chemistry, 2016, pp. 27–42.

[30]   K. I. Salokhiddinov, I. M. Byteva, and B. M. Dzhagarov. Duration of luminescence of singlet oxygen in solution under pulsed laser excitation. *Optika I Spektroskopiya* 47.5 (1979), pp. 881–886.

[31] S. Hackbarth et al. Highly sensitive time resolved singlet oxygen luminescence detection using LEDs as the excitation source. *Laser Physics Letters* 10.12 (2013), p. 125702. DOI: 10.1088/1612-2011/10/12/125702.

[32] A. U. Khan and M. Kasha. Red chemiluminescence of molecular oxygen in aqueous solution. *The Journal of Chemical Physics* 39.8 (1963), pp. 2105–2106. DOI: 10.1063/1.1734588.

[33] L. W. Bader and E. A. Ogryzlo. Reactions of $O_2(^1\Delta_g)$ and $O_2(^1\Sigma_g^+)$. *Discussions of the Faraday Society* 37 (1964), p. 46. DOI: 10.1039/DF9643700046.

PART II

Detection of Photosensitized Generated Singlet Oxygen

CHAPTER 2

Detection Introduction

Due to its photophysical and chemical properties, the generation of $^1O_2$ can be measured using indirect (chemical) or direct (chemiluminescence, phosphorescence) methods. One very important feature in the detection of $^1O_2$ is its long natural lifetime, resulting from the three-fold forbidden transition to the ground state. Consequently, the decay of $^1O_2$ is determined by interaction with its direct environment.

Concerning the intermolecular deactivation of $^1O_2$, two principle pathways have to be distinguished: physical quenching and chemical quenching.

Physical quenching refers to any form of energy transfer to other molecules. In most cases, it consists of only thermal deactivation resulting in dissipating vibrational energy. The rate constants for the quenching reactions can become remarkably high, up to $3 \times 10^9 \, M^{-1} \, s^{-1}$ [1]. Many terminal chemical bonds—first of all O-H, C-H, and N-H—comprise vibrational states that can accept energy from $^1O_2$. The high abundance of such energy acceptors is the reason why $^1O_2$ usually decays much faster than determined by the natural lifetime. In solvents, the decay time of $^1O_2$ has characteristic values, e.g., 3.6 µs for water at room temperature. The bimolecular quenching constants of terminal bonds might not be large (up to $5 \times 10^3 \, M^{-1} \, s^{-1}$), but the effect scales with their abundance, especially if they are part of the solvent molecule.

Chemical quenching means that $^1O_2$ undergoes a chemical reaction with a quencher molecule, forming a new chemical species. Bimolecular rate constants can reach values of up to $1 \times 10^9 \, M^{-1} \, s^{-1}$, e.g., for oxidation of DPBF [2].

Indirect detection methods use the high rate constants of the above-mentioned interactions. Strictly speaking, indirect methods do not observe $^1O_2$, but only the result of its interactions. Most of these methods use absorption, fluorescence, or electron spin resonance (ESR) to determine educts or products of a chemical reaction with $^1O_2$ quantitatively. Historically, these methods originate from their much lower technological threshold for detection compared to the direct detection of $^1O_2$ luminescence. They are quite easy to perform using standard laboratory equipment and are therefore still widely used.

Methods based on physical quenching are less abundant. Photoacoustic detection of the thermal deactivation of $^1O_2$ is neither very specific nor very sensitive and is no longer used today. Recently, delayed fluorescence (DF) of the PS was identified as a useful detection tool under certain conditions. After its generation, $^1O_2$ may encounter some not yet quenched PS molecules in triplet state, resulting in a back transfer of energy to the PS and its subsequent fluorescence. It is claimed, that under certain conditions, DF can be the most sensitive $^1O_2$

detection method [3]. In addition, it is the only indirect method that allows for determination of the decay time of $^1O_2$.

REFERENCES

[1]   S. Mashiko et al. Measurement of rate constants for quenching singlet oxygen with a Cypridina luciferin analog (2-methyl-6-p-methoxyphenyl-3,7-dihydroimidazo 1,2-apyrazin-3-one) and sodium azide. *Journal of Bioluminescence and Chemiluminescence* 6.2 (1991), pp. 69–72. DOI: 10.1002/bio.1170060203.

[2]   G. S. Günther, E. Lemp M, and A. L. Zanocco. Determination of chemical rate constants in singlet molecular oxygen reactions by using 1,4-dimethylnaphthalene endoperoxide. *Journal of Photochemistry and Photobiology A: Chemistry* 151.1-3 (2002), pp. 1–5. DOI: 10.1016/S1010-6030(02)00175-2.

[3]   R. Dědic et al. Parallel fluorescence and phosphorescence monitoring of singlet oxygen photosensitization in rats. *Journal of Innovative Optical Health Sciences* 08.06 (2015), p. 1550037. DOI: 10.1142/S1793545815500376.

CHAPTER 3

Indirect Detection of Singlet Molecular Oxygen

3.1 MONITOR MOLECULES

The functional principle of $^1O_2$ detection via monitor molecules is as follows. The consumption of quencher molecules like DPBF [1] or the formation of reaction products is quantitatively determined after illumination with a known light dose. The concentration of either the sensor molecule or the reaction products thereof is obtained by absorption, fluorescence, or ESR after repeated illumination. Sensitivity and accuracy of this method are limited by the applied detection method and the accuracy of light dose determination.

Observing changes in absorption is the simplest and most accessible method, as it relies on widely available and relatively inexpensive spectrometers.

The indirect method still most commonly adopted is the use of spin traps like tetramethylpiperidine [2], where the product can be quantitatively detected by characteristic ESR signals. This method makes the detection of $^1O_2$ more independent from the optical properties of the sample, the signals are characteristic and usually there is no interference with other signals.

The youngest and most sensitive sensor molecules are switchable linked PET pairs like Singlet Oxygen Sensor Green® (SOSG) [4, 5]. Such pairs consist of a fluorescing molecule and a receptor moiety, in which the fluorophore absorbs and emits light, if possible. 9,10-dimethyl anthracen (DMA) and many derivatives are able to control the fluorescence intensity of a covalently linked xanthene. In the "off" state, electron transfer that deactivates the photoexcited state can successfully compete with the radiative relaxation of the excited molecule (Figure 3.2). Upon chemical reaction of the receptor moiety with $^1O_2$, the PET is deactivated and fluorescence is restored. In the case of DMA, endoperoxides are formed.

In principle, this is a quite powerful detection method and even allows for a certain spatial resolution. The fluorescence intensity of the sensor dyad is several orders of magnitude higher than that of $^1O_2$ phosphorescence. However, several problems related to these sensors limit their applicability, as discussed in Section 3.3.

3.2 DELAYED FLUORESCENCE

Most of the PSs used for generation of $^1O_2$ also have a certain fluorescence rate, which is mostly much bigger than the radiative rate constant of the $^1O_2$ phosphorescence. In addition, fluores-

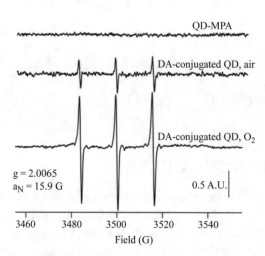

Figure 3.1: Typical ESR signal from 2,2,6,6-Tetramethylpiperidinyloxyl (TEMPO) in toluene (PS: dopamine-conjugated quantum dots) [3].

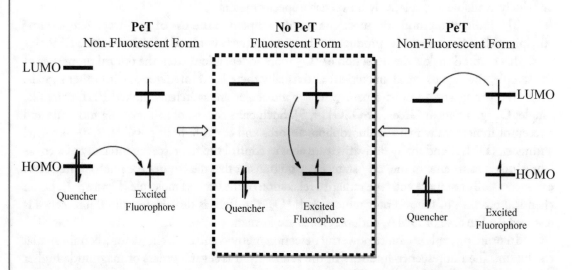

Figure 3.2: Scheme of the photo-induced electron transfer (PET) mechanism in a $^1O_2$ sensor dyad. After excitation of the fluorophore two possible pathways of PET exist, depending on the energetic position of the highest occupied molecular orbital (HOMO) and lowest unoccupied molecular orbital (LUMO) of both moieties. The situation in the middle represents the "on" state, after reaction with $^1O_2$. [4]

$$^3PS^* + {}^3O_2 \rightarrow {}^1PS + {}^1O_2 \qquad \text{1st encounter} - {}^1O_2 \text{ generation}$$

$$^3PS^* \rightarrow {}^1PS^* + {}^3O_2 \quad \text{2nd encounter} - \text{PS back ISC}$$

$$\text{delayed fluorescence}$$

Figure 3.3: Principle of delayed fluorescence for semi-direct $^1O_2$ detection.

cence usually occurs in a wavelength region in which low-cost detectors with high sensitivity can be used. A higher rate constant, lower detector noise, and higher detector efficiency are good reasons to detect the PS fluorescence rather than the $^1O_2$ phosphorescence. In contrast to the initial fluorescence, the delayed PS fluorescence may be quantitatively related to the $^1O_2$ quantum yield under certain circumstances. This method was developed mainly by the groups of Hála [6] and Kubát [7]. After generation during the first encounter with excited photosenitizer in triplet state (^3PS), $^1O_2$ molecules can collide with another excited ^3PS if the PS concentration is sufficiently high. Similar to delayed fluorescence of the E type, the sum of both excitation energies causes the transition of the ^3PS into its singlet excited state (Figure 3.3) from which it can pass over to the ground state via fluorescence. In biologic environments, the probability of fluorescence of an excited PS is 6–7 orders of magnitude higher than that of $^1O_2$ phosphorescence [8, 9].

As a second-order reaction, the energy back transfer from $^1O_2$ to the ^3PS depends on both ^3PS and $^1O_2$ concentration. Therefore, DF detection is the only indirect method that allows to determine the $^1O_2$ kinetics. It must be considered that the $^1O_2$ kinetics is folded with the local PS triplet concentration and thus DF signals exhibit shorter decay times than the real $^1O_2$ kinetics.

Scholz, Dědic, and Hála estimated rate constants for this back transfer in the range of $10^8\,\mathrm{M^{-1}\,s^{-1}}$ for measurements in cells. If the distribution of PS and oxygen would be homogeneous, an initial ^3PS concentration above 10^8 M would result in more DF photons than $^1O_2$ photons.

Similar to all indirect methods of $^1O_2$ detection, the DF suffers from a number of shortcomings that disqualify it for *in vivo* measurements. This method is therefore more of academic interest in this area. In other applications, however, which are associated with lower restrictions in terms of light intensity, light dose and PS selection, it may be useful.

3.3   COMMON LIMITATIONS OF INDIRECT DETECTION

All indirect methods relying on sensor molecules suffer from one major shortcoming: the interaction of interest has to compete with other quenching effects. As long as the sensor molecule can "catch" the $^1O_2$ faster than all other quenchers, the results are reliable. Many $^1O_2$ quantum yields in organic solvents, which were determined this way several years ago, are therefore still valid.

However, the situation is completely different in heterogeneous environments where PS and sensor molecule are not necessarily co-located. When the reaction rate of the sensor molecule becomes diffusion-limited, the obtained results become unreliable. Especially in biological material (e.g., a cell), areas with nearly antithetical properties might be positioned right next to each other. The solubility of PSs and monitor molecules thus may change completely within a few nanometers. Furthermore, for indirect measurement methods to work, the $^1O_2$ has to be able to cover the distance between PS and monitor molecule. In view of the fact that the diffusion length of $^1O_2$ is much shorter than frequently published, the successful indirect detection of $^1O_2$ is even less likely.

In the singlet oxygen community it is often assumed that the average diffusion length would be $\sqrt{2D\tau_\Delta}$ or $\sqrt{6D\tau_\Delta}$, depending on whether a 1D or 3D diffusion in a homogeneous environment is discussed. This estimation falsely implies that all molecules diffuse for a time τ_Δ and decay all at the same moment—which does not reflect the real situation. Due to the exponential decay, the majority of molecules is deactivated before τ_Δ and their displacement will be much less than the σ of the Gaussian distribution. In [10], the diffusion equation for $^1O_2$ with consideration of the decay resulted in a 1D diffusion length of $\sqrt{D\tau_\Delta}$, which is shorter than usually assumed. We will discuss the derivation of this diffusion length more detailed in Subsection 4.2.3.

In the vast majority of cases, considering the 1D diffusion suffices for the question of co-localization as the diffusion length is short compared to the size of biological structures. In the vicinity of membranes, only the diffusion in direction toward the membrane—which a sensor cannot cross—is of interest. Even in water the diffusion length is already below the diffraction limit of optical microscopes. In environments with a high quencher concentration—like in cells—this distance is still shorter. Most of the intracellularly generated $^1O_2$ will never leave the cell. Still, the diffusion length is big enough for a certain percentage of generated $^1O_2$ to pass a membrane, escaping the intracellular monitor molecules. Concerning any quantitative $^1O_2$ detection in a biological environment, this practically rules out all indirect methods.

Furthermore, the accuracy of indirect detection methods relying on the absorption of a sensor molecule cannot be more accurate than absorption itself—which is rather insensitive. Those methods also fail in turbid media or in presence of another strong absorber and many of the sensor molecules tend to form chain reactions.

Methods involving spin traps are limited in sensitivity as well and their accuracy depends on several parameters: PS concentration, illumination dose, and finally the limited signal-to-noise ratio (SNR) of the ESR signal. In order to obtain an interpretable result, several lighting cycles are necessary. In addition, many spin traps are themselves sensitive to illumination, which adds demands toward the spectrum of the excitation light source.

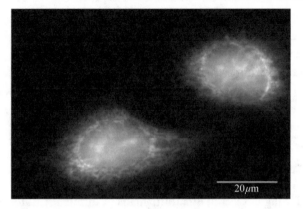

Figure 3.4: HeLa (HeLa) cell incubated with SOSG. The image shows SOSG in the "off" state. SOSG is distributed very inhomogeneously which strongly limits the sensitivity of this detection method. (from [12])

Regarding PET pairs, the most serious problem presents itself with the "off" state of the molecules, which usually is not a dark state. In case of Aarhus Sensor Green (ASG), the fluorescence intensity in the presence of PET is about 10% of the intensity after reaction with $^1O_2$ resulting in a high background signal right from the beginning. The sensitivity is additionally limited by the inhomogeneous distribution of the sensor in structured samples. The variations in signal intensity between the inactive form of PET and the active form caused by the measurement should be higher than the signal variations derived from initial localization. Figure 3.4 illustrates this problem for HeLa cells incubated with SOSG. In the absence of PSs it depicts the sensor in its "off" state equivalent to the situation in $^1O_2$ measurements before any illumination. Even in this state, the signal intensity fluctuates strongly with location, completely independent of any occurrence of $^1O_2$. Sometimes it is difficult to even get the sensor into cells [11], leading to areas with no or low sensor concentration as a consequence and thus making any quantification of the results impossible. Furthermore, some of the sensor molecules themselves generate $^1O_2$ after activation, possibly leading to chain reactions with other sensor molecules and falsifying results even more.

Unlike all other mentioned indirect methods, DF comes along with an intrinsic co-localization of $^1O_2$ generation and sensor molecule (remaining ^3PS) at least shortly after excitation. In case of heterogeneous PS distribution and therefore an inhomogeneous generation of $^1O_2$, the resulting steep concentration gradient causes very fast diffusion of $^1O_2$. Therefore, for a big fraction of $^1O_2$ the remaining ^3PS are soon out of reach, falsifying the determined kinetics and quantum yields. DF also requires a high local excitation intensity, as the probability of the second encounter of ^3PS and $^1O_2$ is directly proportional to the local ^3PS concentration. For the same reason, a highly selective accumulation of the PSs is necessary to achieve this high local ^3PS concentration. The required high excitation intensity can only be achieved in a very small volume. Therefore, illumination limits that apply to biological samples (see Subsection 4.4.1) are easily exceeded before the measurement is finished. This method also requires a reasonable fluorescence quantum yield of the PS, which in turn reduces the $^1O_2$ quantum yield of the PS. Therefore, the efficiency of using DF depends strongly on the PS. Only few PSs have so far been reported to exhibit DF [9] and none of them is even close to clinical application.

> Indirect methods disqualify for any quantitative *in vivo* detection.

REFERENCES

[1]   W. Spiller et al. Singlet oxygen quantum yields of different photosensitizers in polar solvents and micellar solutions. *Journal of Porphyrins and Phthalocyanines* 2.2 (1998), pp. 145–158.

[2]   S. K. Han et al. Evidence of singlet oxygen and hydroxyl radical formation in aqueous goethite suspension using spin-trapping electron paramagnetic resonance (EPR). *Chemosphere* 84.8 (2011), pp. 1095–1101. DOI: 10.1016/j.chemosphere.2011.04.051.

[3]   D. R. Cooper, N. M. Dimitrijevic, and J. L. Nadeau. Photosensitization of CdSe/ZnS QDs and reliability of assays for reactive oxygen species production. *Nanoscale* 2.1 (2009), pp. 114–121. DOI: 10.1039/b9nr00130a.

[4]   K. Krumova and G. Cosa. Fluorogenic probes for imaging reactive oxygen species. *Photochemistry*, ed. by A. Albini and E. Fasani. Vol. 41. Cambridge: Royal Society of Chemistry, 2013, pp. 279–301. DOI: 10.1039/9781849737722-00279.

[5]   C. Flors et al. Imaging the production of singlet oxygen in vivo using a new fluorescent sensor, Singlet Oxygen Sensor Green. *Journal of Experimental Botany* 57.8 (2006), pp. 1725–1734. DOI: 10.1093/jxb/erj181.

[6]   M. Scholz et al. Singlet oxygen-sensitized delayed fluorescence of common water-soluble photosensitizers. *Photochemical & Photobiological Sciences: Official Journal of the European Photochemistry Association and the European Society for Photobiology* 12.10 (2013), pp. 1873–1884. DOI: 10.1039/c3pp50170a.

[7]   J. Suchánek et al. Effect of temperature on photophysical properties of polymeric nanofiber materials with porphyrin photosensitizers. *The Journal of Physical Chemistry B* 118.23 (2014), pp. 6167–6174. DOI: 10.1021/jp5029917.

[8]   S. Hackbarth et al. Time resolved sub-cellular singlet oxygen detection – ensemble measurements versus single cell experiments. *Laser Physics Letters* 9.6 (2012), pp. 474–480. DOI: 10.7452/lapl.201110146.

[9]   M. Scholz, R. Dĕdic, and J. Hála. Microscopic time-resolved imaging of singlet oxygen by delayed fluorescence in living cells. *Photochemical & Photobiological Sciences: Official Journal of the European Photochemistry Association and the European Society for Photobiology* 16.11 (2017), pp. 1643–1653. DOI: 10.1039/c7pp00132k.

[10]   S. Hackbarth, T. Bornhütter, and B. Röder. Singlet oxygen in heterogeneous systems. *Singlet Oxygen*, ed. by S. Nonell and C. Flors. Vol. 2. Comprehensive Series in Photochemical & Photobiological Sciences. The Royal Society of Chemistry, 2016, pp. 27–42.

[11]   Y. Shen et al. Indirect imaging of singlet oxygen generation from a single cell. *Laser Physics Letters* 8.3 (2011), pp. 232–238. DOI: 10.1002/lapl.201010113.

[12]   A. Gollmer et al. Singlet Oxygen Sensor Green (R): photochemical behavior in solution and in a mammalian cell. *Photochemistry and Photobiology* 87.3 (2011), pp. 671–679. DOI: 10.1111/j.1751-1097.2011.00900.x.

CHAPTER 4

Direct Detection of Singlet Molecular Oxygen

The only direct method to verify $^1O_2$ is related to the electronic transition from $^1O_2$ to the triplet ground state, detecting the associated characteristic emission around 1270 nm. Due to the fact that this transition is three-fold forbidden, the radiative rate constant k_r is rather small. For most solvents k_r is smaller than $2\,s^{-1}$ with water having the smallest reported value of $0.11\,s^{-1}$ [1]. Consequently, the local environment, including chemical and physical quenchers therein, determines the $^1O_2$ decay time.

4.1 TECHNICAL ASPECTS

SNR is the major challenge in $^1O_2$ detection: on one hand, the $^1O_2$ phosphorescence is very weak—in water, e.g., just about 1 out of 2.5 million $^1O_2$ molecules emits a photon. On the other hand, the long-wavelength tails of PS fluorescence or phosphorescence in the range of 1270 nm can still be comparatively strong and may interfere with the measurement, even if they are spectrally centered in the UV/VIS or NIR range. Blackbody radiation also plays a role: even if it is comparatively negligible at room temperature, it becomes one of the most important interfering factors some ten degrees above it. In addition, most NIR detectors tend to be noisy by construction.

$^1O_2$ phosphorescence is directly related to the amount of the generated $^1O_2$ and therefore allows for quantitative comparison of different PSs under identical conditions. However, due to the just mentioned issues, discrimination of interfering signals is absolutely necessary for direct $^1O_2$ detection. In principle, both spectral and temporal discrimination is feasible. For historical reasons, steady-state detection is generally combined with spectral discrimination. However, a combination of both discrimination methods is also possible [2]. The differences between the two methods in terms of data acquisition and handling are obvious. In both cases, several high quality commercial solutions are available and commonly in use. Other components, like the excitation sources, light path and detection are more specific to the particular laboratory. However, their impact on the performance of the setup is enormous and will now be discussed.

4.1.1   EXCITATION SOURCES

Apart from the obviously necessary spectral overlap of the excitation with the absorption of the PS, certain requirements concerning focus, modulation, and intensity apply.

Measurements that demand a certain spatial resolution, e.g., 2D scanning or fiber-coupled excitation require at least slight focusing of the excitation light. Despite their broad angular distribution of the emitted light, the small emitting area makes LEDs a possible choice in some cases. However, there usually is barely another option than laser light.

Time-resolved detection by nature requires a modulation of the excitation light, usually pulsed excitation with a steep falling flank. The pulse width should be below 200 ns to allow resolution of typical $^1O_2$ kinetics. In addition to literally pulsed lasers, many cw-laser modules also allow fast modulation and thus qualify as excitation source, resulting in a broader choice of excitation wavelengths. Not all laser light sources are capable of being switched repeatedly or fast enough, though. Furthermore, a deep modulation of the laser is necessary since every background signal decreases the SNR of the setup.

Steady-state detection, in principle, works without any modulation but at the price of very poor SNR, as the signals are weak and the detectors tend to be noisy. To reduce the necessary measurement time, in steady-state detection a higher average excitation intensity than in time-resolved detection is needed. For high-quality steady-state detection, lock-in detection, or similar performant technique is highly recommended. Otherwise, the discrimination of the background noise and other interfering signals from the detected signal is difficult.

All optical components of any setup face certain limitations regarding clear aperture or angle of acceptance. This is most obvious for light that has to be focused into a fiber or to the detector size. We can describe the amount of light that fits both limits (numerical aperture and spot size) of an optical element by the étendue. The overall value of an optical system to guide light from a source to a target is determined by its most limiting component. Whatever additional optics are used, they tend to reduce the usable étendue. In the best case, it can barely be preserved. The practical consequence with regard to the excitation source is that the light, which can be funneled at the intended detection spot or the fiber entrance, has to meet this limitation. An increasing source size always comes at the price of lower usable numerical aperture and vice versa. Despite their broad angular distribution of the emitted light, the small emitting area makes LEDs a possible choice in some cases. In principle, classical light sources can also be used for measurements of $^1O_2$ luminescence, but in addition to their large geometry, the emission often comprises strong near infrared (NIR) emission that complicates discrimination of the excitation light from the $^1O_2$ phosphorescence. Furthermore, switching of thermal light sources is not fast enough, resulting in a poor temporal profile. In most applications, lasers are the light source of choice as their light has low requirements to pass an optical system with low losses due to its low divergence.

4.1.2   COLLECTING OPTICS

As the $^1O_2$ phosphorescence is weak, it is of utmost importance to "collect" emitted light from the biggest possible solid angle around the observation volume. On the one hand, usually the emission is isotropic, so the sensitivity of the setup is directly proportional to the quotient of the collecting solid angle vs. 4π. However, the resulting image size scales with the solid angle of collection as well as with the size of the observation volume. On the other hand, the detector has a limited size and—even more important—a construction-related, limited angle of acceptance. Therefore, the detection image cannot be downscaled at will. Summarized, regarding the collecting optics, the étendue of the complete system is also determined by the weakest component. A good optical pathway is characterized by the fact, that the weakest element is the detector. Following this simple principle directly results in a maximum value for detection volume and observation solid angle and thus best possible sensitivity of the setup.

4.1.3   DETECTORS

Due to the luminescence in the NIR range around 1270 nm and the low $^1O_2$ phosphorescence quantum yield, the detectors for measurement of $^1O_2$ have to meet other requirements than the widely used models for the detection of visible light.

The first detectors used for $^1O_2$ luminescence were Ge-diodes and indium gallium arsenide (InGaAs) diodes, later replaced by avalanche diodes and photomultipliers. Very recently, superconducting nanowires were tested for phosphorescence detection in the NIR.

While the first two detectors can be used in analog mode only, the latter ones were designed mainly for photon-based detection. Presently, all the leading setups worldwide use NIR-Photomultiplier Tubes (PMTs). For time-resolved *in vivo* or *in vitro* measurements there is hardly another option than these PMTs at the moment [3].

Analogue vs. Event-Based Detection

There are three main reasons why counting methods with NIR-PMTs are superior to analog detection. First, time resolution is limited by the detector's response function. The shortest reported value for the 5 −mm Ge PIN diode in the Northcoast Detector EO817P (gold standard for analog $^1O_2$ phosphorescence detection) is approximately 500 ns [4]. Of course, diodes with smaller sensitive area allow for shorter response functions—but at the price of lower sensitivity. In contrast, the transit time spread of a typical NIR-PMT is just about 300 ps. Therefore, when using PMTs it is usually the pulse width of the excitation source, which determines the response function.

Second, any detection system suffers from noise. To quantify the noise of a PIN diode, usually the noise equivalent power (NEP) is used. NEP stands for the minimum power needed to equal the current caused by noise at a given frequency and with a certain bandwidth [5]. The detector itself and the electronic components of the first amplifier stage mainly determine the noise of a photometric system. If need arises for a high bandwidth (fast response), the diodes have to be operated in biased mode and the feedback resistor of the operational amplifier has to be smaller—causing more noise to reach a cut off frequency higher than the one required. Unfortunately, amplifier noise increases with the square of detector capacity hence the size of the sensitive area. In addition to the detector itself, the digital oscilloscope required to record and store the detector voltage increases noise and baseline fluctuations as well.

Counting detectors like single photon avalance diodes (SPADs) or NIR-PMTs face a high probability for thermally induced "false" events. However, this has a relatively low influence as long as sufficient "real" counts are registered in the same time span. Theoretically, the improvement of the SNR is just a matter of measurement time. The random noise follows statistics and equals the square root of the count number, while "real counts" grow linearly. However, limitations may apply if the system under investigation is not stable—a point that is equally valid for analogue detection.

Third, in addition to thermal noise—which is ubiquitous in any detection system—artifacts can falsify the detected signal. These are false, delayed, or otherwise induced signals equally distributed over time or showing a certain distribution pattern in correlation to the excitation. A major source of artifacts in analog detectors is the first amplifier stage after exposure of the PIN diode to the intense long wavelength tail of excitation and the associated interfering signals. The electronics of the preamplifier cannot immediately compensate the big amount of charge induced in the PIN diode at that time. This causes an artifact signal, which for the aforementioned diode appears exactly in the most interesting time range for $^1O_2$ luminescence measurements at 5–20 μs after excitation [6]. In contrast, for event-based detectors it is not the shape of the signal that is important, but only the time of its appearance. Furthermore, the response function of the PMT is sufficiently short so that in most cases single events do not interact with each other. The most important artifact for PMTs is after-pulsing, which mainly depends on the quality and construction of the photocathode and the first dynode [7]. Modern NIR-PMTs like the Hamamatsu H10330-45 show no detectable after-pulsing. Therefore, no detectable artifact attributable to the NIR-PMT is relevant besides the equally distributed dark counts.

Counting devices are superior to analogue devices for any detection of low intensity light.

To illustrate this, the NIR-PMT H10330 from HAMAMATSU was compared to the North Coast detector PO817P at very low signal intensities (Figure 4.1). Pheo in ethanol was strongly diluted to reach signal intensities comparable to those in cells. The Ge-diode requires 30 times more excitation intensity to reach a comparable SNR.

Comparing Detectors for Event-Based Detection
As mentioned before, there are other detector technologies apart from PMTs for single photon detection in the NIR [9]. SPADs and superconducting nanowire single photon detectors (SNSPDs) are discussed in the literature as alternatives to NIR-PMTs.

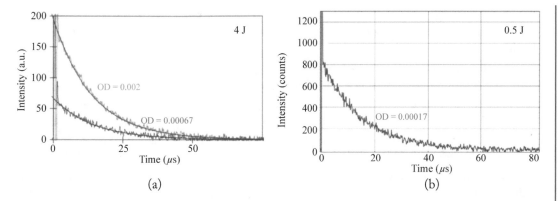

Figure 4.1: Comparison of the singlet oxygen luminescence signal acquired by (A) analog detection with the Northcoast EO817-p and (B) photon counting with the Hamamatsu H10330-45 PMT from solutions of Pheophorbide a (Pheo) in ethanol with different low optical densities. The samples were illuminated with 4 J for analog detection and 0.5 J for photon counting [8].

SPADs are avalanche diodes that are biased above the breakdown voltage to work in Geiger mode. A single electron hole pair, caused by a single photon, is strongly accelerated by the bias voltage and generates additional charges on a large scale, resulting in a detectable current. An occurring avalanche has to be quenched actively by reducing the bias below breakdown voltage. These diodes have a high probability for after-pulsing due to trapped charges. Therefore, the off-time usually extends over several microseconds. While SPADs can generally be operated in gated or free running mode, useful gate intervals are too short to be suitable for $^1O_2$ detection. On the other hand, the free running mode tremendously increases the number of dark counts.

The biggest problem, however, is the long dead time after an avalanche. $^1O_2$ signals depend on many detected photons per cycle, but SPADs soon reach their limit in this use case. With a dead time of a few microseconds, the detector has no chance but to be jammed by an early photon and only become sensitive again after the dead time resulting in an oscillation like pattern. This is especially problematic in cases of strong short time signal artifacts (STAs) at the beginning of each measurement.

SNSPDs are based on superconducting nanowires which are few nanometers thick and about 100 nm wide. In order to achieve a larger detector size, the wire meanders across a wafer to form a bigger detection area. These detectors exploit the existence of a critical temperature and a critical current in superconductors and are conditioned to be close to both limits. Individual infrared photons have enough energy to disrupt hundreds of Cooper pairs in a superconductor, thereby forming a hotspot [10]. This event does not yet block the whole nanowire, but the current avoids the area of hot electrons, driving the local current in the remaining cross section of the nanowire above the limit which causes a complete breakdown of superconductivity.

SNSPDs have a short dead time and nearly no dark counts. The highest reported values of detection efficiency are near 85% [11]. All this makes SNSPDs the perfect detector for $^1O_2$ phosphorescence.

Unfortunately, there is a knock out criterion: SNSPDs have a maximum size of only a few $10\,\mu m$ and a limited aperture due to the direct fiber coupling. However, for dot-like emission sources like in microscopes, such detectors are an option. Even though at least three microscopes with integrated $^1O_2$ phosphorescence detection exist [12, 13], these setups are not useful for biological samples (compare Sections 4.4 and 5.1) apart from the academic fascination. These detectors also work beyond the application in microscopes, but they cannot surpass NIR-PMTs since bigger detector sizes are needed in such experiments. At the moment it is not possible to predict whether it will one day be possible to produce nanowire detectors with bigger sensitive areas.

In principle, steady-state and time-resolved detection have very similar requirements toward the detector. However, the disadvantages of analogue detectors are less pronounced in case of steady-state detection. Artifacts and the longer response time are of less importance here. In many cases, it is therefore possible to reuse already existing equipment for steady-state measurements. In any case, counting detectors should be operated in Geiger mode only, despite the fact that many of them allow analogue operation as well.

4.1.4   STEADY-STATE VS. TIME-RESOLVED DETECTION

To identify $^1O_2$ phosphorescence using steady-state detection one requires spectral discrimination. However, even in solution a quantitative determination of $^1O_2$ without time-resolved measurement is error-prone. The quantification of $^1O_2$ in steady state measurements relies on the fact that in most cases the solvent almost exclusively determines the radiative rate constant of $^1O_2$. Thus, if one compares two samples under identical conditions, the detected signal should be proportional to the amount of generated $^1O_2$. Unfortunately, it is not that simple. The most obvious error source is the PS itself. If it quenches $^1O_2$, the resulting signal will scale with the relative change of τ_Δ. This is especially of importance if the $^1O_2$ decay in the quencher-free solvent is long. In deuterated solvents or in chloroform even weak quenchers may result in completely wrong results. To illustrate this: To achieve the same relative signal reduction in water as in chloroform, the quencher must be about 50 times more concentrated.

It gets even more complicated if the PS is not just a small molecule but embedded in a nanoparticle. If the oxygen can diffuse into the nanoparticle, $^1O_2$ faces two different environments which usually have different decay times and different radiative rate constants, in and outside the nanoparticle. This cannot be detected by measuring the stationary state and it becomes clear that even very simple samples make it difficult to detect the stationary state. It might be useful for monitoring of more- or less-known PSs, but aside from few cases, the method of choice is time-resolved detection. Therefore, we limit the discussion to such measurements.

4.2   FUNDAMENTAL CONSIDERATIONS CONCERNING TIME-RESOLVED DETECTION

Like all luminescence processes, the $^1O_2$ luminescence is also strongly dependent on the microenvironment of all involved components. Since the generation of $^1O_2$ is a multistep bimolecular process, more factors have to be taken into account than for single- or multistep monomolecular processes.

4.2.1   SINGLET OXYGEN LUMINESCENCE KINETICS IN HOMOGENEOUS SYSTEMS

Regarding the generation of $^1O_2$ in a simple environment without diffusional impact, the different $^1O_2$ quenching effects sum up to a single rate constant k_{OG}. In this environment, only three rate-based processes determine the $^1O_2$ kinetics: The Dexter transfer rate k_{TO} from ^3PS to O_2, the sum of all competing processes that deactivate ^3PS without generating $^1O_2$ (k_{TG}) and the sum of all $^1O_2$ quenching processes with k_{OG}.

Solving the corresponding differential equation system results in:

$$\left[^1O_2\right](t) = \left[PS^*\right]_{t=0} \cdot \Phi_\Delta \cdot \frac{k_{TG} + k_{TO}}{k_{OG} - (k_{TG} + k_{TO})} \cdot \left(e^{-(k_{TG}+k_{TO})t} - e^{-k_{OG}t}\right), \quad (4.1)$$

where $[PS^*]_{t=0}$ is the local concentration of excited PS right after excitation.

Since experimental setups usually determine decay times, this equation is often written in the form of

$$\left[^1O_2\right](t) = \left[PS^*\right]_{t=0} \cdot \Phi_\Delta \cdot \frac{\tau_\Delta}{\tau_\Delta - \tau_T} \cdot \left(e^{-t/\tau_\Delta} - e^{-t/\tau_T}\right) \quad (4.2)$$

with the experimentally ascertainable parameters of PS triplet decay time τ_T and $^1O_2$ decay time τ_Δ.

> The two decay times, which describe the $^1O_2$ kinetics in homogeneous environment, may switch their influence on the shape of the $^1O_2$ kinetics. It is always the shorter of the two times that determines the increase, and the longer one that determines the tail.

In aerated solvents like water, the PS triplet decay is usually much faster than the $^1O_2$ decay. The ratio between both times can be influenced by local oxygen partial pressure, diffusion and the presence of quencher molecules. Consequently, the $^1O_2$ kinetics of an unknown sample are not sufficient to assign the specific decay times to the processes involved. This requires independent additional measurements (e.g., ns-transient absorption spectroscopy (TAS) or time-resolved phosphorescence detection of the PS).

4.2.2   INTERFERING SIGNALS

Since the generation of $^1O_2$ depends on the triplet state of a PS, it is not surprising that PS phosphorescence can be an interfering factor in the detection of $^1O_2$. Multiple known PSs emit a strong phosphorescence in the far red and NIR range of the spectrum. For efficient PSs, phosphorescence usually occurs between 1.9 eV (Tetrakismethylpiridiniumylporphine (TMPyP)) and 1.15 eV (Zn phthalocyanines). Lower values may result in energy back transfer to the triplet state of the PS which reduces the $^1O_2$ quantum yield [14]. The corresponding spectral positions of PS phosphorescence maxima are in the region of 700–1100 nm. However, the long wavelength tail of these emissions may overlap significantly with the $^1O_2$ phosphorescence. As the radiative rate constant of $^1O_2$ is comparatively small, PS phosphorescence has to be considered when detecting $^1O_2$ luminescence.

Consequently, $^1O_2$ kinetics should always be fitted with a model that factors in the PS phosphorescence (C in Equation 4.3). BG represents the time-independent background noise.

$$[^1O_2](t) = [PS^*]_{t=0} \cdot \Phi_\Delta \cdot \frac{\tau_\Delta}{\tau_\Delta - \tau_T} \cdot \left(e^{-t/\tau_\Delta} - e^{-t/\tau_T}\right) + C \cdot e^{-t/\tau_T} + BG. \qquad (4.3)$$

The dilemma is obvious: in the attempt to describe the kinetics more accurately, we introduce new parameters, giving the fit even more degrees of freedom. Thus, any parameter that can be determined externally helps to increase the quality of data analysis.

$^1O_2$ measurements also suffer from the ever-present STA. Conceivable reasons for such signals include effects such as scattering or the long-wavelength tail of an intrinsic or PS-related fluorescence or phosphorescence. However, the signals are quite divers. Observations in many different samples give rise to the assumption that the artifact is more pronounced and longer in turbid media, especially in combination with radiationless deactivation of the PS. In contrast, measurements of the same PS in homogeneous solution do not show artifacts with significant intensity in the same set-ups.

STAs have been reported by different groups with different set-ups [13, 15, 16]. Experimental-mathematical elimination of the interfering signal has already been attempted by several groups using measurements with band-pass filters centered at a wavelength adjacent to the one used for $^1O_2$ detection [17, 18]. Such methods can only work if either the disturbing signal has only one component or the spectral distribution of the involved components are known. Otherwise, it may be difficult to unambiguously quantify the disturbing signal (kinetics) at the $^1O_2$ wavelength with just one or two reference measurements.

In the past, several sources of disturbing signals in $^1O_2$ measurements have been identified. These include laser excitation related signals or the luminescence of optical components and sample holders [19, 20]. Such sources of signal distortion can be avoided for current set-ups, because the components can be individually tested for such a luminescence and high-quality quartz-based optics are readily available, even for optical fibers [21].

If the luminescence around the $^1O_2$ wavelength can be measured time-resolved as well as spectrally resolved with sufficient SNR, a procedure similar to decay-associated fluorescence spectra (DAFS) [22] may become applicable to the weak NIR luminescence. However, such a method will still require a parameter-based description of the kinetics of the STA. Since the initial process (excitation) is much shorter than the artifact discussed here, a multi-exponential decay appears to be the logical choice. Even though the exact nature of the signal source is still unknown, it likely comprises one or more probability or rate-based processes. Experience shows that in many cases a double exponential decay is a good choice to describe the STA. At least, due to the given SNR and the small number of signal channels during the first 0.5 μs to 2 μs—which is the typical time range in which this artefact occurs—experience shows that the inclusion of further parameters barely results in higher statistical significance.

4.2.3   INFLUENCE OF DIFFUSION

Up to this point, the generation and detection of $^1O_2$ was only discussed in systems that are not influenced by diffusion effects. However, $^1O_2$ results from energy transfer from a ^{3}PS to molecular oxygen, a process that is strictly limited by diffusion. Following Smoluchowski [23], the corresponding rate constant k_D can be described as:

$$k_D = \gamma \cdot \pi \cdot N_a \cdot 2000 \cdot \left(R_{PS} + R_{O_2}\right)\left(D_{PS} + D_{O_2}\right) \cdot [O_2], \tag{4.4}$$

where R (in m) and D (in $m^2\,s^{-1}$) describe the radius of the interaction cross section and the diffusion constant of the PS and O_2, respectively, while $[O_2]$ (in $mol\,L^{-1}$) is the concentration of O_2.

Strictly spoken, the expanse of the wave function of the excited π^* electron determines the effective radius for the interaction of the ^{3}PS. In case of porphyrins, this is more or less the central ring of the molecule. The ^{3}PS diffusion, however, is determined by its van der Waals surface. Sterical hindering through side groups of the ring and local gradients of the Gibbs enthalpy for oxygen have to be taken into account, as they can modify the local O_2 concentration. Each of these influences can be represented by a constant factor; these factors can be combined into a positive constant γ, which is usually smaller than one.

> The rate of $^1O_2$ generation is mainly determined by the local concentration and diffusion rate of oxygen. However, the structure and conformation of the PS may alter the generation rate.

Quenching of $^1O_2$ is nearly exclusively diffusion limited and mainly happens in the microsecond time range. The preceding processes—PS excitation and ISC—can be considered instantaneous on this time scale.

Apart from some special cases, where the $^1O_2$ source and quenchers are directly coupled to each other, the decay of $^1O_2$ is primarily determined by diffusion limited quenching. The natural radiative decay of $^1O_2$ is neglectable as it is too slow to compete with quenching processes.

> Independent of the environment, the radiative deactivation has almost no influence on the $^1O_2$ kinetics.

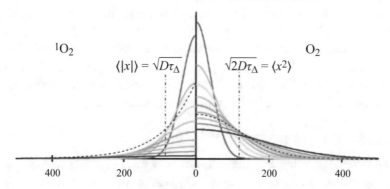

Figure 4.2: Distribution of $^1O_2$ and O_2 after diffusion at times, when multiples of 10% of the $^1O_2$ decayed. For $^1O_2$ the normalized overall average of the diffusion length (dashed curve) and for O_2 the average diffusion length after τ_Δ (dashed curve) are indicated.

In addition to these "obvious" processes, diffusion of $^1O_2$ between regions with different radiative rate constants also modifies the detected phosphorescence kinetics. After generation, $^1O_2$ will diffuse depending on the local environment. Whether this diffusion has consequences for the determined $^1O_2$ phosphorescence kinetics depends on the composition of the local environment, more precisely: on the size of volumes in which parameters differ from those of the surrounding areas. As long as these areas with different properties are either small or big compared to the diffusion length of $^1O_2$, heterogeneity does not need to be addressed separately. In such cases the signal is a superposition or a weighted average of signals from different homogeneous regions.

In all environments where aforementioned limiting cases do not apply, the diffusion of singlet oxygen has a significant effect on the luminescence kinetics. $^1O_2$ is soluble in many materials (from gases to solids) and it is quite small and therefore very mobile. It is assumed that the diffusion constants of $^1O_2$ are comparable to those of molecular oxygen since size and charge of the molecule do not change [24]. This assumption is supported by experimental evidence for similar partition coefficients between water and hydrophobic media for molecular oxygen and $^1O_2$ [24]. However, in contrast to molecular oxygen, $^1O_2$ exhibits a lot of interactions with its surroundings. Therefore, the average diffusion length is quite short, much shorter than stated in most publications on that topic. To explain this in more detail, the distribution of molecular oxygen and $^1O_2$ in water shall be compared. While the amount of oxygen is constant, the one of $^1O_2$ decays exponentially with τ_Δ.

The curves in Figure 4.2 represent the spatial distribution of $^1O_2$ (left) and molecular oxygen (right) after 1D diffusion when 10–90% of the initially generated $^1O_2$ decayed. On the right side, the dashed black curve represents the distribution after τ_Δ. The standard deviation σ of this distribution is described by $\sqrt{2D\tau_\Delta}$.

This value is often falsely associated with the average diffusion length of $^1O_2$, which would mean that all $^1O_2$ molecules diffuse for a time τ_Δ and decay at the same moment. This does not reflect the real situation. Due to the exponential decay, the majority of molecules is already deactivated before τ_Δ and their displacement is likely much shorter than σ of the Gaussian distribution.

The representation of σ by $< x^2 >$ instead of $< x >$ in the Gaussian distribution over-simplifies the description of diffusion processes. In reality, the average displacement is rather $< |x| >$ than $< x^2 >$. Of course, this also includes the task of taking the direction of the displacement into account. The question of how many molecules can reach a quencher molecule or leave a membrane, however, cannot be answered by the average squared distance in any case.

The dashed curve on the left side of Figure 4.2 is the result of the integration of the $^1O_2$ distribution curve over time and its subsequent normalization. This corresponds to the spatial distribution of $^1O_2$ deactivation and thus the distribution of diffusion lengths of the $^1O_2$ molecules. It is described by an exponential function:

$$\frac{1}{2\sqrt{D\tau_\Delta}} \exp \frac{-|x|}{\sqrt{D\tau_\Delta}}, \; x \in (-\inf, \inf). \tag{4.5}$$

To calculate this distribution and the average diffusion length, in [16] we suggested to include the decay of the singlet oxygen into the 1D-diffusion equation:

$$\frac{\partial C(x,t)}{\partial t} = D\frac{\partial^2 C(x,t)}{\partial x^2} - k_{OG}C(x,t), \tag{4.6}$$

leading to the solution of:

$$C(\widetilde{x},t) = \frac{1}{2\sqrt{\pi t}} \exp -\frac{\widetilde{x}^2}{4t} - k_{OG}t \tag{4.7}$$

with $\widetilde{x} = x/\sqrt{D}$.

Equation (4.7) represents a Gaussian distribution that decreases and widens over time. For this solution of the extended diffusion equation, the x axis needs to be scaled by $1/\sqrt{D}$. Averaging and rescaling of the x axis then results in the above-mentioned distribution of diffusion lengths with an average diffusion length of $\sqrt{D\tau_\Delta}$ for $^1O_2$.

The average free path of $^1O_2$ in 1D diffusion is $\sqrt{D\tau_\Delta}$.[a] In water this value is only 85 nm; in the presence of quenchers it is even lower.

[a]A very similar situation in semiconductor physcis—the diffusion length of minority charge carriers—is described identically.

Considering three-dimensional diffusion, this coefficient of course would be $\sqrt{3D\tau_\Delta}$. This only comes into play when the investigated structures are smaller than the diffusion range of $^1O_2$.

In practice, most questions can be answered by a one-dimensional consideration of the systems, as is the case for the question whether $^1O_2$ can diffuse through a membrane.

However, if a higher percentage of $^1O_2$ molecules diffuses across a phase border, the signal kinetics of the $^1O_2$ phosphorescence will be affected. The main reason for this influence of diffusion on the kinetics are the radiating rate constants, which vary depending on the material. With respect to the shape of the kinetics, the diffusion of $^1O_2$ from an area with a higher radiative rate constant has the same effect as an additional decay channel for $^1O_2$ in this area. The signal decays faster than it would without diffusion and cannot be compensated by amplification in adjacent regions where the radiative rate constant is lower.

> Any quantitative $^1O_2$ phosphorescence measurement in heterogeneous (e.g., biological) environments has to account for diffusion and different radiative rate constants.

4.3   DETECTION IN SYSTEMS OF HIGHER COMPLEXITY

While the detection of $^1O_2$ luminescence and its theoretical treatment in homogeneous systems is relatively simple, more complex systems present a much greater challenge and make the interpretation of the resulting signals much more difficult. In practice, measurements on complex biological systems such as cell suspensions and biofilms become of increasing importance, but the difficulties already arise in much simpler model systems such as micelles. The consideration of superpositions from different areas of these samples increases the degrees of freedom while the significantly lower signal intensity reduces the SNR. In the following sections, we will shed light on two of the most common cases of more complex systems.

4.3.1   PHOTOSENSITIZING NANOPARTICLES

Many applications comprise PSs in form of nanoparticles (NPs), whether for targeted delivery or easy removal after the application. Either the PSs are embedded in a matrix or the NP itself can generate $^1O_2$, as is the case for quantum dots or NPs of polymeric semiconductors. As a consequence, $^1O_2$ is generated mainly inside the NP and due to the size of the NP, only a fraction of the generated $^1O_2$ will diffuse far enough to interact with the surroundings of the NP. This presents the additional challenge of distinguishing between the two fractions of $^1O_2$, which might be of interest for application in PDT. The further away from the surface $^1O_2$ is generated, the smaller its chance to reach the surface of the NP by diffusion before being quenched. Consequently, the larger the NPs, the higher the fraction of $^1O_2$ phosphorescence originating from inside of it. In the likely case of different radiative rate constants of $^1O_2$ in and outside the NP, the signal from inside the NP is over- or underpronounced in the overall signal. However, the resulting deviation of the $^1O_2$ kinetics from the normal double-exponential kinetics may allow to separate $^1O_2$ phosphorecsence according to its origin.

The $^1O_2$ kinetics in NPs and solvent are different, since both are affected by $^1O_2$ diffusion in an opposite manner: A gain of $^1O_2$ concentration on one side of the system is a loss on the other. Additionally, these changes are accompanied by a change of signal intensity due to the different radiative rate constant in solvent and NP.

For small NPs (few nanometers) the $^1O_2$ signal from inside the NP decays with τ_T, since the diffusion outwards is comparatively fast. As long as τ_Δ in the solvent is longer, the $^1O_2$ phosphorescence originating from solvent and NP can easily be distinguished [2]. For bigger NPs, numerical simulations will be necessary to analyze the $^1O_2$ phosphorescence kinetics.

It is quite obvious that with steady state detection one would not be able to distinguish $^1O_2$ phosphorescence from in- and outside the NP, falsifying the result. In this scenario it would of course be possible to supplement steady-state detection with the indirect detection using sensor molecules. Those sensor molecules would identify only the $^1O_2$ leaving the NP thus giving correction values for the $^1O_2$ measured via steady-state detection. Unfortunately, sensor molecules quench $^1O_2$ outside the NP during their detection process. This will increase the concentration gradient of $^1O_2$ between the interior of the NP and the outside and increase the diffusion, yielding too big values.

4.3.2   PHOTOSENSITZERS IN MEMBRANES

In [25] it was shown that the $^1O_2$ phosphorescence kinetics of a membrane-localized PS differs from that of a water-soluble one. It turned out that compared to the surrounding water, the $^1O_2$ phosphorescence coming from membranes is over pronounced by a factor of about 10 due to higher oxygen solubility and the radiative rate constant of $^1O_2$ [25, 26]. Consequently, quantitative comparison of different PSs in biological environment is rather difficult, even using time-resolved detection.

To illustrate this, imagine a membrane with a much longer $^1O_2$ lifetime than the $^1O_2$ lifetime in the surrounding medium and both being longer than the PS triplet decay time. All $^1O_2$ is generated inside the membrane following the PS triplet decay time. $^1O_2$ starts to diffuse out immediately after generation and it takes just a few nanoseconds for the first molecules to leave the membrane but soon the concentration gradient driving the process flattens. After the diffusion stabilizes the fraction quotient of $^1O_2$ concentration at the interface, all $^1O_2$ molecules, even in the membrane, follow the $^1O_2$ decay outside.

To describe the phosphorescence coming from such a system, diffusion cannot be neglected like in the bi-exponential model for homogeneous environments (Equation (7.1)). As shortly after excitation the majority of $^1O_2$ is located inside the membrane, where the radiative rate constant of $^1O_2$ is higher than in the medium, the $^1O_2$ luminescence will deviate from the bi-exponential model (see Figure 4.3). Fitting kinetics from such systems with the standard bi-exponential decay model will result in poor fits and wrong results. Since the $^1O_2$ kinetics in heterogeneous systems can vary strongly, independent verification of the PS triplet decay time becomes even more important.

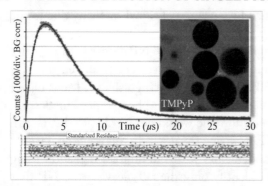 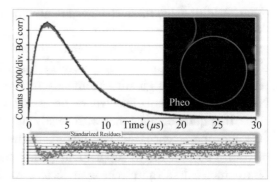

Figure 4.3: With $^1O_2$ phosphorescence kinetics it is possible to identify the localization of PS. The water-soluble TMPyP generates $^1O_2$ phosphorescence kinetics typical for water, while the membrane-localized PS Pheo causes deviations of the $^1O_2$ phosphorescence kinetics from the standard double exponential model. The inserts show Confocal Laser Scanning Microscopy (CLSM) images of giant unilamellar vesicles (GUVs) with the PS (raw data) that confirm the positioning at the lipid bilayer [25].

As already mentioned in Subsection 4.3.1, steady-state detection is not sufficient to identify the influence of sample heterogeneity but only presents the overall signal, which is highly affected by $^1O_2$ diffusion and the effects related to it.

> For heterogeneous samples like *in vivo* systems, the only advisable option for singlet oxygen detection is the time-resolved measurement of its phosphorescence.

4.4   OBJECTIVE DETECTION LIMITS

In Section 4.1.3, we explained why analog devices cannot compete with counting methods. Therefore, in this chapter we will only cover the latter. Note that for analog detection, in addition to the limitations listed for counting detectors in the following, bandwidth limits or drift effects and electronic artifacts further reduce accuracy.

4.4.1   LIMITS DETERMINED BY SAMPLE STABILITY

As mentioned in Section 4.1.3, counting detection can theoretically reach any aspired SNR if the measurement time can be extended accordingly. $^1O_2$, however, is highly reactive and therefore many samples under investigation—especially *in vivo*—change during a measurement. Chemical quenching consumes oxygen and modifies the quencher, leading to a change in both the $^1O_2$ and PS triplet decay time. Several parameters influence this effect: the quencher concentration, the quenching constant, oxygen supply, oxygen solubility, and diffusion in the sample.

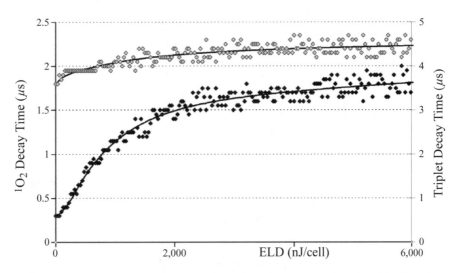

Figure 4.4: Development of t_T (grey) and t_D (black) of Pheo in LNCaP (LNCaP) cell suspension under continuous illumination [27].

In Figure 4.4, the development of $^1O_2$ and PS triplet decay times is shown for ongoing illumination of a typical membrane-localized PS in cell suspensions [26, 27]. Due to the chemical quenching, both parameters initially increase very rapidly after the start of illumination. While strong stirring can partially compensate the oxygen consumption in suspension, the oxygen supply in voluminous samples can soon become the most important parameter influencing $^1O_2$ kinetics.

After prolonged illumination, both parameters converge to a final value. The PS triplet time approaches a stationary state, where the oxygen consumption equals the amount of oxygen moving in from outside. The increase of the $^1O_2$ decay slows when chemical quenchers are nearly depleted and physical quenching determines the $^1O_2$ decay exclusively.

Therefore, $^1O_2$ phosphorescence measurements in biological material always require a certain detection volume (number of cells investigated). We will discuss this particular environment more detailed in Chapter 5.

4.4.2   LIMITS DETERMINED BY PS STABILITY AND DETECTOR DARK COUNTS

If the PS bleaches after excitation, the usable measurement time is limited. This applies in particular to signals that initially have lower count rates than the dark noise of the detection system. While the additional gain of the signal counts decreases with duration of the measurement, the noise added by the dark counts retains and after a certain time leads to an annulment of the SNR gain. The effect is shown in Figure 4.5 for exponentially bleached PSs with three different

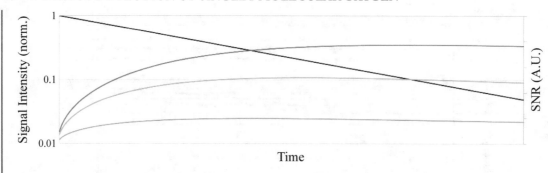

Figure 4.5: Development of SNR for bleaching PS with therefor exponentially decreasing intensity (black), shown for starting count rates five times the dark noise (red), equal to the dark noise (green) and one fifth of the dark noise (blue).

initial count rates: five times the noise, equal to the noise and one fifth of the noise. In the latter case, with regard to SNR, it does not pay to continue the measurement after the PSs bleached to half of their intensity.

> The maximum possible gain in SNR due to prolonged measurement duration is limited by PS bleaching.

Another consequence is that the shorter the measurement time, the better the SNR for a given illumination dose. The shorter the total acquisition time, the lower the detector noise and the lower the illumination required to achieve a certain SNR. Of course, this only applies as long as no other limits apply, e.g., detector artifacts or oxygen supply.

As could be shown with few selected examples, time-resolved detection of $^1O_2$ phosphorescence is the only applicable method, when the local environment around the PS becomes increasingly complex. Radiative rate constants of $^1O_2$ can be very diverse in the different phases of the heterogeneous sample structure and $^1O_2$ diffusing between phases results in different kinetics in either phase. If the local geometry of the phases near the PS is known, time-resolved $^1O_2$ phosphorescence detection allows a rather accurate description even for *in vitro* or *ex vivo* systems.

However, what we call a "real *in vivo* measurement" inside living organisms or biofilms results in a more complex description, which is particularly complicated by the local heterogeneity of such samples (Chapters 6 and 7).

REFERENCES

[1] T. D. Poulsen, P. R. Ogilby, and K. V. Mikkelsen. Solvent effects on the $O_2(a^1\Delta_g)$–$O_2(X^3\Sigma_g^-)$ radiative transition: comments regarding charge-transfer interactions. *The Journal of Physical Chemistry A* 102.48 (1998), pp. 9829–9832. DOI: 10.1021/jp982567w.

[2] S. Pfitzner et al. Time- and spectrally resolved singlet oxygen phosphorescence detection—discriminating disturbance signals. *Laser Physics* 28.8 (2018), p. 085702. DOI: 10.1088/1555-6611/aac5cc.

[3] A. Jiménez-Banzo et al. Time-resolved methods in biophysics. 7. Photon counting vs. analog time-resolved singlet oxygen phosphorescence detection. *Photochemical & Photobiological Sciences: Official Journal of the European Photochemistry Association and the European Society for Photobiology* 7.9 (2008), pp. 1003–1010. DOI: 10.1039/b804333g.

[4] R. Schmidt and C. Tanielian. Time-resolved determination of the quantum yield of singlet oxygen formation by tetraphenylporphine under conditions of very strong quenching †. *The Journal of Physical Chemistry A* 104.14 (2000), pp. 3177–3180. DOI: 10.1021/jp994057n.

[5] N. Bertone and P. Webb. Noise and Stability in PIN Detectors. *EG&G Canada* (1996).

[6] S. Oelckers et al. Time-resolved detection of singlet oxygen luminescence in red-cell ghost suspensions: Concerning a signal component that can be attributed to 1O2 luminescence from the inside of a native membrane. *Journal of Photochemistry and Photobiology B: Biology* 53.1-3 (1999), pp. 121–127. DOI: 10.1016/S1011-1344(99)00137-2.

[7] Hamamatsu Photonics K.K. *Photomultiplier Tubes: Basics and Applications, 3rd Edition*. Hamamatsu, 2007.

[8] J. C. Schlothauer, S. Hackbarth, and B. Röder. First Sino-German Symposium on "Singlet molecular oxygen and photodynamic effects": Current prospects of detectors for high performance time-resolved singlet oxygen luminescence detection. *Photonics & Lasers in Medicine* 4.4 (2015), pp. 303–306. DOI: 10.1515/plm-2015-0032.

[9] G. S. Buller and R. J. Collins. Single-photon detectors for infrared wavelengths in the range 1–1.7 micrometer. *Springer Series on Fluorescence*, Springer Berlin Heidelberg, 2014. DOI: 10.1007/4243_2014_64.

[10] C. M. Natarajan, M. G. Tanner, and R. H. Hadfield. Superconducting nanowire single-photon detectors: physics and applications. *Superconductor Science and Technology* 25.6 (2012), p. 063001. DOI: 10.1088/0953-2048/25/6/063001.

[11] I. E. Zadeh et al. *A Single-photon Detector with High Efficiency and Sub-10ps Time Resolution*. 2018. arXiv: 1801.06574 [physics.ins-det].

[12] S. Hatz, Lambert, John D. C., and P. R. Ogilby. Measuring the lifetime of singlet oxygen in a single cell: addressing the issue of cell viability. *Photochemical & Photobiological Sciences: Official Journal of the European Photochemistry Association and the European Society for Photobiology* 6.10 (2007), pp. 1106–1116. DOI: 10.1039/b707313e.

[13] M. Scholz et al. Real-time luminescence microspectroscopy monitoring of singlet oxygen in individual cells. *Photochemical & Photobiological Sciences: Official Journal of the European Photochemistry Association and the European Society for Photobiology* 13.8 (2013), pp. 1203–1212. DOI: 10.1039/C4PP00121D.

[14] P. A. Firey et al. Silicon naphthalocyanine triplet state and oxygen. A reversible energy-transfer reaction. *Journal of the American Chemical Society* 110.23 (1988), pp. 7626–7630. DOI: 10.1021/ja00231a007.

[15] M. T. Jarvi et al. The influence of oxygen depletion and photosensitizer triplet-state dynamics during photodynamic therapy on accurate singlet oxygen luminescence monitoring and analysis of treatment dose response. *Photochemistry and Photobiology* 87.1 (2011), pp. 223–234. DOI: 10.1111/j.1751-1097.2010.00851.x.

[16] S. Hackbarth, T. Bornhütter, and B. Röder. Singlet oxygen in heterogeneous systems. *Singlet Oxygen*, ed. by S. Nonell and C. Flors. Vol. 2. Comprehensive Series in Photochemical & Photobiological Sciences. The Royal Society of Chemistry, 2016, pp. 27–42.

[17] M. Pfitzner et al. Prospects of in vivo singlet oxygen luminescence monitoring: Kinetics at different locations on living mice. *Photodiagnosis and Photodynamic Therapy* 14 (2016), pp. 204–210. DOI: 10.1016/j.pdpdt.2016.03.002.

[18] M. J. Niedre et al. Measurement of singlet oxygen luminescence from AML5 cells sensitized with ALA-induced PpIX in suspension during photodynamic therapy and correlation with cell viability after treatment. *Optical Methods for Tumor Treatment and Detection: Mechanisms and Techniques in Photodynamic Therapy XI*, ed. by T. J. Dougherty. Vol. 4612. International Society for Optics and Photonics. SPIE, 2002, pp. 93–101. DOI: 10.1117/12.469336.

[19] R. D. Scurlock, K.-K. Iu, and P. R. Ogilby. Luminescence from optical elements commonly used in near-IR spectroscopic studies: The photosensitized formation of singlet molecular oxygen ($^1\Delta_g$) in solution. *Journal of Photochemistry* 37.2 (1987), pp. 247–255. DOI: 10.1016/0047-2670(87)85005-0.

[20] K.-K. Iu and P. R. Ogilby. Formation of singlet molecular oxygen ($^1\Delta_g O_2$) in a solution-phase photosensitized reaction. 2. A comment on static quenching. *The Journal of Physical Chemistry* 92.16 (1988), pp. 4662–4666. DOI: 10.1021/j100327a021.

[21] J. C. Schlothauer et al. Luminescence investigation of photosensitizer distribution in skin: correlation of singlet oxygen kinetics with the microarchitecture of the epidermis. *Journal of Biomedical Optics* 18.11 (2013), p. 115001. DOI: 10.1117/1.JBO.18.11.115001.

[22] J. R. Knutson, D. G. Walbridge, and L. Brand. Decay-associated fluorescence spectra and the heterogeneous emission of alcohol dehydrogenase. *Biochemistry* 21.19 (1982), pp. 4671–4679. DOI: 10.1021/bi00262a024.

[23] M. Smoluchowski. Versuch einer mathematischen Theorie der Koagulationskinetik kolloider Lösungen. *Zeitschrift f. Physik. Chemie* (1917), pp. 129–168.

[24] R. M. Cordeiro. Reactive oxygen species at phospholipid bilayers: Distribution, mobility and permeation. *Biochimica et Biophysica Acta* 1838.1 Pt B (2014), pp. 438–444. DOI: 10.1016/j.bbamem.2013.09.016.

[25] S. Hackbarth and B. Röder. Singlet oxygen luminescence kinetics in heterogeneous environment – Identification of the photosensitizer localization in small unilamellar vesicles. *Photochemical & Photobiological Sciences: Official Journal of the European Photochemistry Association and the European Society for Photobiology* (2014). DOI: 10.1039/C4PP00229F.

[26] S. Hackbarth et al. New insights to primary photodynamic effects–Singlet oxygen kinetics in living cells. *Journal of Photochemistry and Photobiology B: Biology* 98.3 (2010), pp. 173–179. DOI: 10.1016/j.jphotobiol.2009.11.013.

[27] S. Hackbarth et al. Time resolved sub-cellular singlet oxygen detection – ensemble measurements versus single cell experiments. *Laser Physics Letters* 9.6 (2012), pp. 474–480. DOI: 10.7452/lapl.201110146.

PART III

$^1O_2$ in Biological Systems

CHAPTER 5

Time-Resolved Singlet Oxygen Luminescence in Cell Suspensions

In contrast to the systems covered in Part II, cell suspensions, whether eukaryotic or prokaryotic, are much more complex. In the model systems discussed so far, even in the most complex case (liposomes) only three different microenvironments occur in a very small space. In living systems, the natural degree of local disorder is known to be significantly higher, a circumstance that is necessary for the maintenance of life. Eukaryotic cells present either with a cell wall (plant) or a cell membrane (animal). In contrast to liposomes, those consist of a multitude of different phospholipids with different chain lengths, degrees of saturation and head group modifications, as well as numerous proteins embedded in or attached to the membrane. These cell envelopes, which can be regarded as more complex liposomes, have a differentiated compartmentation in their interior, by which cellular processes are spatially separated and which additionally increases the variety of micro-environments. Although prokaryotes have no compartments and cellular processes occur throughout the cell interior, they are no-less complicated from the point of view of $^1O_2$ phosphorescence measurement. One of the factors contributing to the complexity of prokaryotes is the fact that the cell membrane of prokaryotes is protected by a cell wall.

Still, in most biological material water represents the majority of the overall volume. Thus, in most cases it defines an upper limit for the $^1O_2$ decay time. Cellular quenchers such as proteins, nucleic acids, polyunsaturated fatty acids, and cell-expressed radical scavengers like bilirubin further shorten this decay time [1–3]. Among all of these quenchers, proteins are of particular interest; they exhibit high bimolecular reaction constants for oxidation and are essential for cell survival. Oxygen-sensitive proteins may have a fixed location (i.e., membrane-bound) or move freely inside the cell, such as catalase or superoxide dismutase.

The resulting $^1O_2$ decay time in cellular suspensions is thus largely dependent on its generation site, i.e., the localization of the PS. For membrane-affine PSs, despite the highly variable composition of membranes, the reported decay time varies only slightly and is significantly shorter than the upper limit presented by pure water. Water-soluble PSs tend to result in longer decay times much closer to the reference value in water. However, since molecules other than water quench the majority of $^1O_2$ in cells, the decay time is still shorter. When $^1O_2$ is quenched not physically but chemically (most common for quenchers of biological relevance and the basis for a successful PDT/PDI), quencher and oxygen are consumed, resulting in increasing decay times of both $^1O_2$ and PS triplet state.

5.1   MAMMALIAN CELLS

As the first practical application of the photodynamic effect studied by modern medicine is the fight against cancer. The first measurements of $^1O_2$ phosphorescence in biological samples also took place in suspensions of mammalian cells. Several investigators have published estimates of the decay time of $^1O_2$ within cells [4–7]. For membrane-affine PSs most of them reported decay times of 0.5 μs or less.

For water-soluble PSs, slightly longer decay times (≈ 1.5 μs) have been reported [8]. However, molecules other than water quench the vast majority of $^1O_2$ in cells and the majority thereof quench $^1O_2$ chemically. This is also the reason why the photodynamic effect eventually results in cell death. Chemical quenching consumes both the quencher and oxygen. As a direct consequence, both their concentrations decline over time and the $^1O_2$ decay time as well as the PS triplet decay time gets longer. Maisch et al. first observed the influence of the local oxygen concentration on the PS triplet time in suspensions of Staphylococcus aureus (*S. aureus*) [9]. Albeit being measured in bacterial cells instead of eukaryotic cells, this influence was later validated for Jurkat cells incubated with Pheo [6, 10].

Schlothauer, Hackbarth, and Röder (2009) were the first to observe a change in $^1O_2$ kinetics in cell suspensions during successive measurements. Both decay times involved tend toward higher values with increasing illumination, which could be observed for membrane localized PSs in different cell suspensions [6, 11]. The initial $^1O_2$ decay time in not illuminated cells amounts to 0.4 μs and increases drastically during illumination with only 1 μJ per cell, much faster than it does in a homogenous solution in the presence of quenchers (see Subsection 4.4.1). After an illumination with 7 μJ per cell, the $^1O_2$ decay time reaches a value of 1.5 μs and is further rising under continued illumination. At this point, however, 50% of the cell samples proofed to be non-viable. The PS triplet decay time also rises during illumination, indicating an additional consumption of oxygen due to the chemical quenching of $^1O_2$.

Already in 2006, the group of Ogilby conducted first experiments to determine the $^1O_2$ kinetics in single cells. Using a single-photon microscope and a light dose of 6 mJ per cell, the group succeeded in determining the $^1O_2$ decay time to 3.2 μs [12]. Since at a light dose of 7 μJ per cell 50% of the cells of a sample are already no longer alive (see above), it must be assumed that measurements of $^1O_2$ kinetics in single cells can only be equivalent to results after cell death (see Subsection 4.4.1). This assumption is confirmed by the measured decay time that correlates with the decay time of $^1O_2$ in water. It can therefore be concluded that this value represents the upper limit of the $^1O_2$ decay time in biological samples, which is caused by pure physical quenching of the $^1O_2$ luminescence by water as well as the cell components remaining after chemical quenching. While the determination of $^1O_2$ kinetics in single cells is therefore not feasible, those experiments remain to be of great value to the community.

Combining the results of Schlothauer, Hackbarth, and Röder and Ogilby, the process of PDT inside a cell can be described as follows: Based on an initial $^1O_2$ decay time of 0.4 µs in cells without illumination, about 85% of all $^1O_2$ is quenched by cellular components at the beginning the PDT. The proportion of available biological quenchers decreases drastically with increasing illumination during a successful PDT until the $^1O_2$ decay time rises to a value of 1.5 µs. At this point, only 50% of the $^1O_2$ is still chemically quenched by the molecules present in a cell sample. Further illumination still leads to an increase in $^1O_2$ decay time; however, due to the reaction radius of $^1O_2$ and diffusion constraints, this development happens much slower than during the initial phase of the photosensitization.

Weston and Patterson addressed the question, whether this development of the parameters can be better described by fixed or flowing chemical quenchers. Using a Monte Carlo simulation they demonstrated that both PS and quenchers are likely fixed in the cellular environment during the above-mentioned experiment. While at the beginning the chemical quenchers where arranged in a grid around the PS, this grid becomes pitted with time and the $^1O_2$ decays slower. The parameters of the simulation could be adapted to retrace the experimental results. However, the authors had to limit the final $^1O_2$ lifetime to 1.2 µs to get a good match with the experimental data. That is much shorter than the aforementioned upper limit of around 3 µs, probably because the simulations of Weston and Patterson were based on a single type of quencher, either fixed or free flowing. All published results, including those of single cell experiments, can only be explained by a combination of both quencher types. Unlike fixed quenchers, diffusing quenchers can be replaced from all over the cytoplasm. Under these conditions, the corresponding changes in lifetime occur much slower. On the spatial scale the simulation was run in, the influence of flowing quenchers is rather small. Therefore, the final lifetime as assumed by Weston and Patterson [13] is determined by physical and floating chemical quenchers. Further illumination then only slowly increases the $^1O_2$ decay time until—after much higher illumination doses—finally there are just physical quenchers left.

Consequently, membrane-bound as well as hydrophilic PSs, face chemical quenchers in cells. However, quenching efficiency, kinetics, and therefore impact on the cell vitality will be different depending on their localization.

To illustrate this, Jurkat cells were briefly incubated for just 30 min with two PSs similar in structure but with different hydrophobicity: Pheo and Chlorin e6 (Ce6). The $^1O_2$ phosphorescence kinetics were correlated to the phototoxic effect after low-dose illumination with white light LEDs (60 mJ cm^{-2}, Figure 5.1).

For the selected concentrations, the overall $^1O_2$ phosphorescence signal of Pheo was much smaller than the one of Ce6, even though similar amounts of $^1O_2$ were generated in both cases. Analysis of the signal shows that this effect can be solely attributed to the much shorter $^1O_2$ decay time of Pheo compared to the one of Ce6. For Pheo, which localizes near the inner membranes of the cells, the interaction of $^1O_2$ with cellular quenchers is obviously much stronger. Therefore, the PDT efficiency of Pheo on cells after short incubation times is much higher than

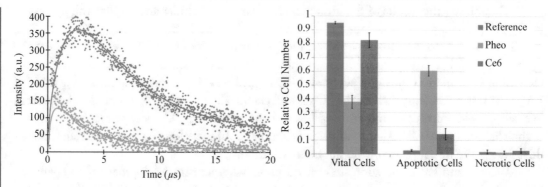

Figure 5.1: While the intensity of the $^1O_2$ luminescence signal generated by Ce6 is larger than the one generated by Pheo after incubation of Jurkat cells for 30 min, the treatment impact is lower for incubation with Ce6. This can probably be attributed to the lower interaction of $^1O_2$ with cellular quenchers due to different localization of $^1O_2$ generation, resulting in a longer $^1O_2$ decay time for samples incubated with Ce6 compared to Pheo [14].

that of Ce6. The cell vitality test 90 min after illumination supports this interpretation (Figure 5.1 right). While the majority of cells is apoptotic after incubation with Pheo, the majority is still alive after incubation with Ce6 [14]. This is consistent with other reports that PSs achieve the highest efficiency near cell membranes [15].

> For *in vitro* samples, time-resolved detection of $^1O_2$ is necessary to obtain useful results, as in most cases the effect of $^1O_2$ is of interest. However, it is precisely this effect that shortens the $^1O_2$ decay, reducing the overall signal intensity and falsifying the result of steady-state measurements.

5.2   MICROORGANISMS

While the basic principle of $^1O_2$ phosphorescence in bacteria corresponds to that in eukaryotic cells, new experimental challenges arise solely from the difference in size of the organisms studied. Thus, it is not surprising that the first measurements of the kinetics of $^1O_2$ luminescence on bacteria were only published in 2007 [9].

In addition to the fundamentally different cellular structure, bacteria (with the exception of individual cases) are on average 10 times smaller than eukaryotic cells. This difference in size leads to the fact that the luminescence signals of $^1O_2$ produced by PSs in bacteria often either have a significantly lower amplitude or the scattering of the samples superimposes the signal of $^1O_2$ phosphorescence. Those effects occur if the same volume cell number as in eukaryotic samples is used or when the cell number is increased to reflect the biological mass of eukaryotic samples, respectively.

However, due to the occurrence of both aerobic and anaerobic metabolisms in bacteria and shorter reproduction cycles, it is much easier to investigate the influence of O_2 concentration on $^1O_2$ kinetics than is the case for basically aerobic eukaryotic cells. Using *S. aureus* incubated with photofrin, Maisch et al. [9] succeeded for the first time in demonstrating that the triplet decay time of a PS in bacteria increases considerably with decreasing O_2 concentration. The upper limit for this increase is the natural triplet lifetime of the PS in the absence of oxygen. As is also the case during PDT on eukaryotic cells, a successful PDI leads both to a reduction of the quencher concentration of a sample and to a reduction of the O_2 concentration. As expected, the rate of this decrease is strongly dependent on the biological mass initially contained in the samples. This is to be expected, as it is the interaction between $^1O_2$ and the biological material that leads to a decrease in both concentrations.

With Photofrin, a PS known from PDT was used in aforementioned experiment and Maisch et al. had to find out that this was not taken up by gram negative bacteria according to fluorescence microscopic images. It is already known today that the PSs used for PDT are not suitable for use against bacteria due to the different structure of the cell wall. As Ragàs et al. published in 2010 [16], the meanwhile widely known PS TMPyP can, however, be detected in samples of the gram negative Escherischia coli (*E. coli*). Although gram-negative and gram-positive bacteria differ strongly, the rise and decay times measured in both experiments were similar and lay between the expected 3 µs and 14 µs for water or lipids, respectively.

In these as well as in the following published results it was therefore wrongly assumed for a long time that $^1O_2$ is produced in membranes as well as in the aqueous environment of the samples. The differences to the development of kinetics in eukaryotic cells were attributed to the different size of the actual cells, according to which a larger oxygen gradient is achieved within eukaryotic cells than in bacterial cells, but bacterial conglomerates can have larger oxygen gradients than individual cells. This interpretation of the results was justified by the state of knowledge and technology at that time and could only be conclusively refuted by Müller, Preuß, and Röder in 2018 due to considerably improved detection techniques.

As early as 2013, Preuß et al. indicated by analyzing fluorescence decay times that TMPyP primarily attaches itself to the exterior of bacterial cells and can only penetrate them as soon as a photodynamic effect has occurred or as soon as membrane permeability-increasing substances are used. Due to the small size of bacterial cells, unlike eukaryotic cells, the exact localization of PS in bacteria cannot be determined directly under the microscope.

By analyzing the kinetics of TMPyP in samples of two *E. coli* strains, Müller, Preuß, and Röder were able to confirm once again that PS is not exposed to any altered microenvironment until the photodynamic destruction of the cells begins. For this purpose, not only the fluorescence spectrum but also the $^1O_2$ kinetics of samples after different incubation times were compared with PS both in the dark and under illumination. The results show that even after 120 min of incubation in the dark, the spectral shape of the fluorescence and the $^1O_2$ luminescence kinetics correspond exactly to those after the addition of PS. In contrast, incubation under illumination showed a bathochromic shift of fluorescence indicating a changed microenvironment of the PS. However, a much stronger indication is the change in the $^1O_2$ luminescence kinetics, which can no longer be described by the usual double exponential term after only 30 min of incubation but is obviously composed of several superimposed signals from different microenvironments. As Müller, Preuß, and Röder could show, the temporal course of the changing $^1O_2$ luminescence kinetics also corresponds to the course of the phototoxicity on the samples.

For the determination of the $^1O_2$ luminescence kinetics in microorganisms a significantly higher SNR than for the determination in eukaryotic cells is necessary. With sufficient sensitivity, the $^1O_2$ luminescence kinetics can provide information about the position of a PS in relation to the bacterial cell.

REFERENCES

[1]   R. W. Redmond and I. E. Kochevar. Spatially resolved cellular responses to singlet oxygen. *Photochemistry and Photobiology* 82.5 (2006), pp. 1178–1186. DOI: 10.1562/2006-04-14-IR-874.

[2]   A. P. Castano, T. N. Demidova, and M. R. Hamblin. Mechanisms in photodynamic therapy: part one-photosensitizers, photochemistry and cellular localization. *Photodiagnosis and Photodynamic Therapy* 1.4 (2004), pp. 279–293. DOI: 10.1016/S1572-1000(05)00007-4.

[3]   R. Kondo et al. Identification of heat shock protein 32 (Hsp32) as a novel survival factor and therapeutic target in neoplastic mast cells. *Blood* 110.2 (2007), pp. 661–669. DOI: 10.1182/blood-2006-10-054411.

[4]   M. T. Jarvi et al. The influence of oxygen depletion and photosensitizer triplet-state dynamics during photodynamic therapy on accurate singlet oxygen luminescence monitoring and analysis of treatment dose response. *Photochemistry and Photobiology* 87.1 (2011), pp. 223–234. DOI: 10.1111/j.1751-1097.2010.00851.x.

[5]   J. W. Snyder et al. The imaging of singlet oxygen in single cells. *Proc. SPIE 5689*, ed. by SPIE. Vol. 5689. 2005, pp. 17–25. DOI: 10.1117/12.590239.

[6]   S. Hackbarth et al. New insights to primary photodynamic effects–Singlet oxygen kinetics in living cells. *Journal of Photochemistry and Photobiology B: Biology* 98.3 (2010), pp. 173–179. DOI: 10.1016/j.jphotobiol.2009.11.013.

[7]   A. Baker and J. R. Kanofsky. Quenching of singlet oxygen by biomolecules from L1210 leukemia cells. *Photochemistry and Photobiology* 55.4 (1992), pp. 523–528. DOI: 10.1111/j.1751-1097.1992.tb04273.x.

[8] A. Jiménez-Banzo et al. Kinetics of singlet oxygen photosensitization in human skin fibroblasts. *Free Radical Biology & Medicine* 44.11 (2008), pp. 1926–1934. DOI: 10.1016/j.freeradbiomed.2008.02.011.

[9] T. Maisch et al. The role of singlet oxygen and oxygen concentration in photodynamic inactivation of bacteria. *Proceedings of the National Academy of Sciences of the United States of America* 104.17 (2007), pp. 7223–7228. DOI: 10.1073/pnas.0611328104.

[10] J. C. Schlothauer, S. Hackbarth, and B. Röder. A new benchmark for time-resolved detection of singlet oxygen luminescence – revealing the evolution of lifetime in living cells with low dose illumination. *Laser Physics Letters* 6.3 (2009), pp. 216–221. DOI: 10.1002/lapl.200810116.

[11] S. Hackbarth et al. Time resolved sub-cellular singlet oxygen detection – ensemble measurements versus single cell experiments. *Laser Physics Letters* 9.6 (2012), pp. 474–480. DOI: 10.7452/lapl.201110146.

[12] J. W. Snyder et al. Optical detection of singlet oxygen from single cells. *Physical Chemistry Chemical Physics* 8.37 (2006), pp. 4280–4293. DOI: 10.1039/b609070m.

[13] M. A. Weston and M. S. Patterson. Effect of 1O2 quencher depletion on the efficiency of photodynamic therapy. *Photochemical & Photobiological Sciences: Official Journal of the European Photochemistry Association and the European Society for Photobiology* 13.1 (2014), pp. 112–121. DOI: 10.1039/c3pp50258a.

[14] S. Hackbarth et al. First Sino-German Symposium on "Singlet molecular oxygen and photodynamic effects": Time-resolved singlet oxygen luminescence detection under PDT-relevant light doses. *Photonics & Lasers in Medicine* 4.4 (2015), pp. 299–301. DOI: doi:10.1515/plm-2015-0032. URL: https://doi.org/10.1515/plm-2015-0032.

[15] I. O. L. Bacellar et al. Membrane damage efficiency of phenothiazinium photosensitizers. *Photochemistry and Photobiology* 90.4 (2014), pp. 801–813. DOI: 10.1111/php.12264.

[16] X. Ragàs et al. Cationic porphycenes as potential photosensitizers for antimicrobial photodynamic therapy. *Journal of Medicinal Chemistry* 53.21 (2010), pp. 7796–7803. DOI: 10.1021/jm1009555.

[17] A. Müller, A. Preuß, and B. Röder. Photodynamic inactivation of Escherichia coli - Correlation of singlet oxygen kinetics and phototoxicity. *Journal of Photochemistry and Photobiology B: Biology* 178 (2018), pp. 219–227. DOI: 10.1016/j.jphotobiol.2017.11.017.

[18] A. Preuß et al. Photoinactivation of Escherichia coli (SURE2) without intracellular uptake of the photosensitizer. *Journal of Applied Microbiology* 114.1 (2013), pp. 36–43. DOI: 10.1111/jam.12018.

CHAPTER 6

Time-Resolved Singlet Oxygen Luminescence Detection in Microorganisms on Surfaces

The photodynamic inactivation of microorganisms is, in comparison to other fields of $^1O_2$ deployment, a relatively new one. Being proposed shortly after discovery of the photodynamic effect, it was discontinued when the synthesis of new antimicrobial compounds such as antibiotics seemed to make its investigation irrelevant. Only after the development of multiple resistances in bacteria in an alarmingly short time, the topic began to resurface in the early 1990s. The parallel investigation of biofilms, the natural sessile growth form of all microorganisms, revealed, that the pathogenic potential of bacteria only arises when forming those complex colonies on surfaces. The increased resistance of microorganisms in biofilms against antimicrobial agents and their change in metabolism and gene expression form a necessity for the monitoring of $^1O_2$ in microorganisms growing on surfaces.

Contrary to direct $^1O_2$ luminescence detection in solutions or suspensions, the time-resolved detection of $^1O_2$ on surfaces poses a much higher challenge both to the measurement setup as well as its sensitivity [1]. This is especially true when those surfaces are covered with microorganisms or biofilms thereof. The increased requirements for measurements of microorganisms on surfaces not only include the possibility of scanning a 2D-surface with an acceptable resolution [1]. Additionally, those systems add complexity to the measurements due to interfering signals, lower concentration of the $^1O_2$ producing agents, lower concentration of the microorganisms, and the none-planar surface of naturally growing biofilms [1–4]. The low concentration of the microorganisms on the surfaces in comparison to their concentration in suspensions also leads to higher light doses for every single cell while at the same time producing much lower signal intensities.

Aforementioned difficulties make it understandable, why the majority of investigations into the PDI of biofilms doesn't use direct time-resolved $^1O_2$ luminescence measurements but confines to indirect measurements or verification of $^1O_2$ generation prior to PDI [4, 5].

6.1   BACTERIA

The PDI of bacteria was proven to be an effective antibacterial method, however, this only holds true while their cells grow in the planktonic state. If allowed to grow as a biofilm, they are reported to put up much more of a fight. The results of PDI on bacterial biofilm varies not only between different photosensitizers, concentrations, and illumination doses but also between the cellular makeup of the biofilm [6, 7]. At that, Maisch et al. [8] established that the phototoxicity is directly affected by the $^1O_2$ quantum yield of the utilized photosensitizer, but it is not the only property that is influencing the reaction of the biological system within the biofilm. SAPYR, a photosensitizer with an exceptional high quantum yield of 99%, can kill bacteria in monospecies biofilms at a 5-fold higher rate than the well-known TMPyP [2]. Furthermore, it can disrupt the formation of the extracellular biofilm matrix due to its tenside character. But even photosensitizers like methylene blue, which is known to have low $^1O_2$ quantum yield and photostability, can reduce the bacterial count in dental biofilms if only illuminated with a high enough dose of light [9].

Yet, in all these systems, the photosensitizer is applied in suspension and the $^1O_2$ kinetic inside the biofilm samples wasn't measured. Still, a photosensitizer immobilized on a surface under the right conditions can generate $^1O_2$ [5]. Analysis of the $^1O_2$ kinetic of these surface samples without microorganisms showed, that the generated $^1O_2$ can diffuse out of the surface. An indirect measurement with potassium iodide confirmed the results of the time-resolved detection of $^1O_2$ phosphorescence. A porphyrin-doped polymer coating was able to reduce the cfu on the surfaces by three log steps. A similar result was achieved by Müller, Preuß, and Röder [4] using electron beam functionalized membranes with TMPyP. Here, a 2D-measurement of $^1O_2$ kinetics of the membranes proved the evenly distributed functionalization of the membranes. For E. coli deposited onto these membranes by pipetting of a bacterial suspension, a reduction of six log levels could be demonstrated for illuminated membranes in comparison to reference samples.

A successful direct time-resolved detection of $^1O_2$ generated during PDI of bacteria on surfaces was performed by Bornhütter [1]. For this study, a wild type stem of E. coli was used that was previously shown to be photodynamically inactivated by the photosensitizers TMPyP and the cationic corrole PCor+. The complexity of the measurements in this system is enhanced by interfering signals from the normal cultivation media for bacteria as well as the conflict between the time consuming measurements and the relatively short generation time of bacteria. Furthermore, the study showed that bacteria itself are vastly reducing the $^1O_2$ luminescence signal at high concentrations. While the research investigated the $^1O_2$ luminescence signals on samples over a period of nine days, the results show, that a $^1O_2$ luminescence signal on samples incubated with bacteria could only be detected for a period of up to two days. Through isolation of the $^1O_2$ luminescence signals and various background signals, Bornhütter [1] was able to monitor a change in kinetics for one of the photosensitizers in correlation to incubation time, initial bacteria concentration and growth of the bacteria on the samples.

The time-resolved measurement of $^1O_2$ phosphorescence of photodynamically inhibited bacterial biofilms on surfaces during PDI is still pending.

6.2   FUNGI

Fungi prove much harder to kill than bacteria, be it using biocides or PDI, due to their higher resistance against environmental conditions. Their susceptibility toward PDI varies greatly between different species but, in general, much higher light doses are needed to reach the same reduction as is achieved for bacteria. The well-known cationic photosensitizer TMPyP is still able to reduce the amount of planktonic cells and biofilms of Candida albicans by five and six log steps, respectively [10]. Novel corrole-PS like PCor+ are even found capable to inactivate the spores of mold fungi [11]. If it holds true that fungi need higher illumination doses than bacteria for a successful PDI, it is even more true, if they grow in biofilms. The measurement of time resolved $^1O_2$ phosphorescence under these conditions show an unusual kinetic for TMPyP during the PDI with no clearly detectable rise time of the signal.

Preuß et al. [12] were able to detect 2D time-resolved $^1O_2$ phosphorescence of TMPyP incorporated into façade paints before and during inactivation of the mold fungi Aspergillus niger and Cladosporium cladosporoides. Comparison of the kinetics shows that similar to reports of Felgenträger et al. [13], stacking of PS in fungi leads to changed luminescence signal. This was observed for all microorganisms but is most distinct in fungi, presumably because of high intracellular availability of iron.

Bornhütter et al. [14] also succeeded to measure $^1O_2$ luminescence kinetics on surfaces under PDI relevant conditions of dermatophytes and molds. Using Trichophyton rubrum and Scopulariopsis brevicaulis with TMPyP and PCor+ as photosensitizers, they were able to monitor the $^1O_2$ generation and its kinetic spatially resolved over a period of seven days. Using a highly sensitive setup [1, 14], they show that the $^1O_2$ kinetic inside the fungi slowly changes from a bi-exponential form toward a monoexponential form as seen in [10] and [12]. The speed of this signal transformation strongly depends on the species of fungus and can be correlated to the interaction of the photosensitizer with the fungi cells. Additionally, both the triplet state lifetime and the $^1O_2$ lifetime change during the PDI of fungi. However, it is not yet known why this happens and how it corresponds to the phototoxicity.

6.3   PHOTOTROPHIC MICROORGANISMS

The PDI of phototrophic microorganisms is far less prominent than that of bacteria and fungi. This can most likely be attributed to the lack of danger generally posed by algae and cyanobacteria. Indeed, one of the first attempts at photodynamic inactivation of phototrophic microorganisms concerns the containment of cyanobacterial blooms. Drábková, Maršálek, and Admiraal [15] investigated the effect of anionic photosensitizers and methylene blue on cyanobacteria and green algae in order to find a measure that affects the former more than the latter. The results showed no efficient PDI, though in 2008 a project named BIODAM took up the concept to prevent damage to cultural heritage, which is also mostly caused by phototrophic microorganisms. Using a combination of cell permeabilizers, biocides, pigment inhibitors, and PDI, they succeeded in reducing biofilm growth of green algae [16]. The share of PDI to that result was low though, as methylene blue and nuclear fast red were used as photosensitizers and only worked in combination with H$_2$O$_2$.

Better results are achieved using photostable cationic photosensitizers like TMPyP and PCor+ [17] which result in an inhibition over 18 days without depending on permanent illumination. 2D scans of the development of ^{1}O$_2$ kinetics during the PDI of green algae show a strong change of the signal over the time [3]. In contrast to measurements of pure photosensitizer, the signal intensity decayed much faster in contact with algae cells and the signal formed changed similar to the experience on bacteria and fungi from a bi-exponential signal to a mono-exponential one. Bornhütter et al. [3] especially showed the importance of direct detection of ^{1}O$_2$ is, since phototrophic microorganisms themselves exhibit a strong fluorescence signal and the distribution of photosensitizers cannot be monitored using fluorescence alone.

From the few published results presented here, it becomes obvious that the measurement of direct ^{1}O$_2$ luminescence kinetics on surfaces with microorganisms is still a difficult topic. The main obstacles stem from the low luminescence signals and the isolation of interfering signals from the ^{1}O$_2$ luminescence. It is still impossible to determine, in detail, how the presence of microorganisms influences the ^{1}O$_2$ luminescence signal in these systems and how the differences in kinetics during the PDI might be explained. Nevertheless, the fact that such highly sensitive measurements were unthinkable of only a few years ago in both spatial as well as temporal regime let alone in the presence of microorganisms proof, that this part of the ^{1}O$_2$ luminescence detection is one to pay close attention to in the coming years.

REFERENCES

[1] T. Bornhütter. Nutzung der orts- und zeitaufgelösten Detektion der Singulettsauerstoff Lumineszenz zur Evaluierung der Photodynamischen Inaktivierung von Mikroorganismen. PhD thesis. Berlin, Germany: MatNatFak, Humboldt-Universität zu Berlin, 2018. DOI: 10.18452/19124.

[2] F. Cieplik et al. Photodynamic biofilm inactivation by SAPYR–an exclusive singlet oxygen photosensitizer. *Free Radical Biology & Medicine* 65 (2013), pp. 477–487. DOI: 10. 1016/j.freeradbiomed.2013.07.031.

[3] T. Bornhütter et al. Development of singlet oxygen luminescence kinetics during the photodynamic inactivation of green algae. *Molecules* 21.4 (2016), p. 485. DOI: 10.3390/ molecules21040485.

[4] A. Müller, A. Preuß, and B. Röder. Photodynamic inactivation of Escherichia coli - Correlation of singlet oxygen kinetics and phototoxicity. *Journal of Photochemistry and Photobiology B: Biology* 178 (2018), pp. 219–227. DOI: 10.1016/j.jphotobiol.2017.11.017.

[5] A. Felgenträger et al. Singlet oxygen generation in porphyrin-doped polymeric surface coating enables antimicrobial effects on Staphylococcus aureus. *Physical Chemistry Chemical Physics* 16 (2014), pp. 20598–20607.

[6] C. R. Fontana et al. The antibacterial effect of photodynamic therapy in dental plaque-derived biofilms. *Journal of Periodontal Research* 44 (2009), pp. 751–759.

[7] A. Kishen et al. Efflux pump inhibitor potentiates antimicrobial photodynamic inactivation of Enterococcus faecalis biofilm. *Photochemistry and Photobiology* 86 (2010), pp. 1343–1349.

[8] T. Maisch et al. The role of singlet oxygen and oxygen concentration in photodynamic inactivation of bacteria. *Proceedings of the National Academy of Sciences of the United States of America* 104.17 (2007), pp. 7223–7228. DOI: 10.1073/pnas.0611328104.

[9] K. M. de Freitas-Pontes et al. Photosensitization of in vitro biofilms formed on denture base resin. *The Journal of Prosthetic Dentistry* 112 (2014), pp. 632–637.

[10] F. P. Gonzales et al. Fungicidal photodynamic effect of a twofold positively charged porphyrin against Candida albicans planktonic cells and biofilms. *Future Microbiology* 8 (2013), pp. 785–797.

[11] A. Preuß et al. Photodynamic inactivation of mold fungi spores by newly developed charged corroles. *Journal of Photochemistry and Photobiology B: Biology* (2014), pp. 39–46. DOI: 10.1016/j.jphotobiol.2014.02.013.

[12] A. Preuß et al. Photodynamic inactivation of biofilm building microorganisms by photoactive facade paints. *Journal of Photochemistry and Photobiology B: Biology* (2016). DOI: 10.1016/j.jphotobiol.2016.04.008.

[13] A. Felgenträger et al. Ion-induced stacking of photosensitizer molecules can remarkably affect the luminescence detection of singlet oxygen in Candida albicans cells. *Journal of Biomedical Optics* 18 (2013), p. 04502.

[14] T. Bornhütter et al. Singlet oxygen luminescence kinetics under PDI relevant conditions of pathogenic dermatophytes and molds. *Journal of Photochemistry and Photobiology B: Biology* (2017). DOI: 10.1016/j.jphotobiol.2017.12.015.

[15] M. Drábková, B. Maršálek, and W. Admiraal. Photodynamic therapy against cyanobacteria. *Environmental Toxycology* 22 (2007), pp. 112–115.

[16] M. E. Young et al. Development of a biocidal treatment regime to inhibit biological growths on cultural heritage: BIODAM. *Environmental Geology* 56 (2008), pp. 631–641.

[17] J. Pohl et al. Inhibition of green algae growth by corrole-based photosensitizers. *Journal of Applied Microbiology* 118 (2015), pp. 305–312.

CHAPTER 7

Time-Resolved Singlet Oxygen Luminescence *Ex Vivo* and *In Vivo*

Direct $^1O_2$ luminescence detection is one of the few possibilities to undoubtedly detect $^1O_2$ without further intervention in the system under investigation. Under *in vivo* conditions most indirect methods will fail, especially those that rely on any kind of reporter substances. Furthermore, time resolution is desired in order to detect the presence of $^1O_2$ on the one hand and to gain insights in the process of $^1O_2$ generation and quenching on the other hand. As already mentioned, a kinetics analysis of the time-resolved $^1O_2$ signal at 1270 nm can provide knowledge about the PS triplet decay time, which is strongly correlated with oxygen availability. Additionally, the $^1O_2$ lifetime, which is dominated by quenching, could provide a measure for the efficacy of the therapy.

With these advantages, the reason for the great interest in $^1O_2$ luminescence detection *in vivo* and especially during PDT becomes obvious. The $^1O_2$ luminescence detection for dosimetry purposes is a common goal [1–4]. Unfortunately, the challenges of detecting the extremely weak NIR signals are even more severe when investigating *in vivo* samples compared to *in vitro* which is already much more challenging then PSs in homogeneous environments. Not only the detection but also the analysis of the data is much more difficult due to the very inhomogeneous nature of *in vivo* samples as well as various possible sources of NIR luminescence in the investigated spectral region. Especially the complications connected to the detection are a major reason for the manageable amount of literature. Nevertheless, in certain model systems, detection of $^1O_2$ and sometimes even an analysis of the kinetics was possible.

The following models are commonly used to describe $^1O_2$ luminescence kinetics. In these models τ_Δ is the $^1O_2$ lifetime, τ_T the PS triplet lifetime A a positive amplitude proportional to the $^1O_2$ quantum yield (Φ_Δ) and O an offset caused by detector dark counts:

$$I(t) = \frac{A}{1 - \frac{\tau_T}{\tau_\Delta}} \left(e^{-\frac{t}{\tau_\Delta}} - e^{-\frac{t}{\tau_T}} \right) + O. \tag{7.1}$$

Equation (7.1) is valid only for homogeneously distributed PSs but is often used to describe *in vitro* or *in vivo* kinetics as well. Equation (7.2) expands this model with one or more exponential terms. In case of one additional term and $\tau_n \equiv \tau_T$, a PS phosphorescence contribution can be described:

$$I(t) = \frac{A}{1 - \frac{\tau_T}{\tau_\Delta}} \left(e^{-\frac{t}{\tau_\Delta}} - e^{-\frac{t}{\tau_T}} \right) + \sum_0^N B_n \cdot e^{-\frac{t}{\tau_n}} + O. \qquad (7.2)$$

Due to the usually moderate SNR of *in vivo* NIR signals it is important to minimize the amount of free parameters for fitting. In contrast to measurements in solution, the STA (see Subsection 4.2.2) with a typical decay time of $\tau_1 \leq 1\,\mu s$ has a much higher influence on the data. Since a phosphorescence signal can be observed for many PSs it might be necessary to consider an additional exponentially decaying term [5–8]. The PS phosphorescence usually increases at low oxygen levels. Due to the size of the detection area, a superposition of slightly different oxygen levels is likely to be observed while PSs in areas with low oxygen availability dominate the measured PS phosphorescence.

Therefore, Equation (7.2) with $N = 1$ or $N = 2$ is a good approach to fit *in vivo* NIR data. The obtained decay times correspond to the PS triplet decay time of PSs effectively generating $^1O_2$ (τ_T), $^1O_2$ lifetime (τ_Δ), the STA (τ_1), and the PS triplet decay time of PSs in a micro environment with lower oxygen saturation (τ_2).

7.1   3D CELL CULTURE

An *ex vivo* model system closest to *in vivo* environments are 3D cell cultures. Cells grow in a collagen matrix and form cell clusters. Nutrients, oxygen, and PS can diffuse from the surrounding culture medium into the cluster where they are taken up by the cells.

Fluorescence images of a 3D cell culture with FaDu (ACC-784) (FaDu) using mTHPC as PS are shown in Figure 7.1. The PS accumulates in the cells and no fluorescence is visible from the surrounding matrix as can be seen from the light microscopy image on the left side. Furthermore, a localization of the PS in cytomembranes can be observed from the CLSM image (right side).

These 3D cell cultures can be scanned for NIR luminescence, but with a much lower resolution. Exemplary kinetics are shown in Figure 7.2 together with a fit according to Equation (7.2) (N=2), the 96% confidence interval and the estimated $^1O_2$ proportion.

The offset caused by dark counts should be about 2.5 whereas a clear deviation is visible in the data. Measurements performed at a detection wavelength of 1200 nm show this ostensible high offset as well. Therefore, it is actually a very long decaying signal, namely the PS phosphorescence in areas with a low oxygen saturation. The fit yields a $^1O_2$ decay time of $\tau_\Delta = (1.0 \pm 0.2)\,\mu s$ a triplet decay time for PS that efficiently generate $^1O_2$ of $\tau_T = (7.2 \pm 0.8)\,\mu s$ a fast-decaying disturbance signal with $\tau_1 = (0.3 \pm 0.1)\,\mu s$ and a long-decaying PS phosphorescence with $\tau_2 \approx (320 \pm 40)\,\mu s$.

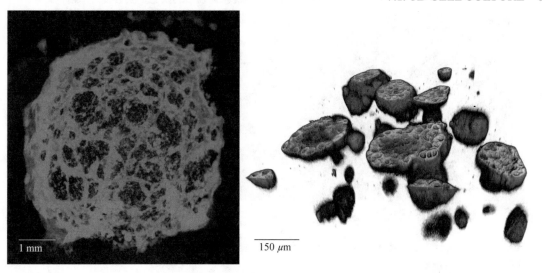

Figure 7.1: Fluorescence light microscopy image (left, $\approx$ 5 mm) and CLSM image (right, $\approx$ 300 μm edge length) of a 3D cell culture, dyed with mTHPC grown in Cultrex®.

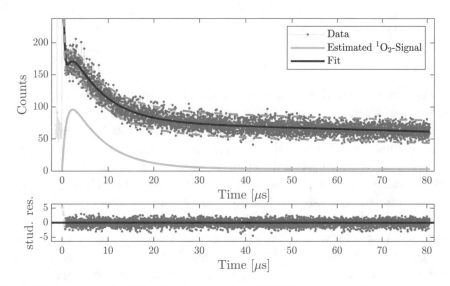

Figure 7.2: NIR kinetics of a 3D cell culture with mTHPC at 1270 nm [8].

Even though the cell culture was covered with phosphate-buffered solution (PBS) and pure oxygen was blown onto it before and during the measurement, that long-decaying PS phosphorescence dominates the NIR signal. Since the 3D cell culture is a rather large object (about 5 mm in diameter and several 100 μm in height) oxygen supply is strongly limited by diffusion.

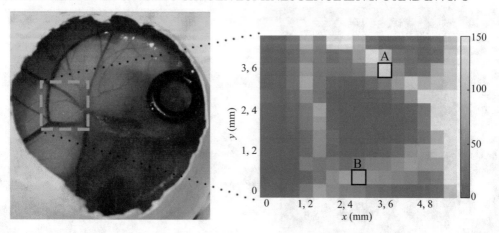

Figure 7.3: Image of CAM with arteries and veins. Scanned area is marked and $^1O_2$ amplitude is shown on the right [8].

Compared to cell suspensions the volume ratio between cells with PS and surrounding medium without PS, acting as oxygen reservoir, is much lower wherefore the strong PS phosphorescence is not observed in the cell suspensions.

7.2 CAM

The first *in vivo* model to be discussed here that allows $^1O_2$ luminescence detection with high SNR and a detailed kinetics analysis is the chorioallantoic membrane (CAM) model. This well-established scientific model is widely used for research on transplantation, metastasis (anti-) angiogenesis, and others [9–13]. Even though *in vivo* investigations are possible, it is not considered to be an animal experiments and therefore accessible to physicists.

In this model system, one can easily investigate PSs in blood vessels, which is of particular interest when treating skin conditions like port wine stain [14]. Blood vessels, supplying the tumor with nutrients and oxygen, are also of great interest in tumor therapy [15–17].

Usually the vascular system in the CAM is well developed and contains vessels with a diameter large enough for injections after Egg Development Day (EDD) nine [18]. It is possible to grow tumors on the CAM and observe $^1O_2$ kinetics [19–22]. Furthermore, veins and arteries, which differ greatly in their oxygen partial pressures, are located beneath the surface of the CAM. Since the PS triplet decay time depends strongly on oxygen availability, significantly different $^1O_2$ luminescence kinetics are to be expected.

Figure 7.3 shows a photograph of an egg 240 min after PS injection on EDD 11 with some large blood vessels. The area indicated by the green rectangle was scanned for $^1O_2$-luminescence kinetics. In this area, an artery (low blood oxygen saturation) on the left and bottom side and a vein (high blood oxygen saturation) on the top-right side can be seen. To avoid $^1O_2$ generation

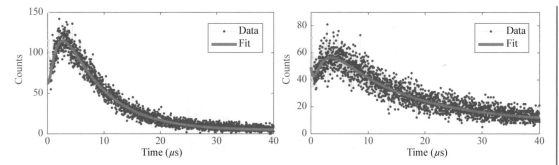

Figure 7.4: Different kinetics from veins (left) and arteries (right) measured at pixel A and B, respectively [8].

and therefore photodynamic actions, yellow light was used during sample preparation. In these lighting conditions, it is not possible to see the vastly different colors of the veins and arteries.

The data obtained in each point was fitted according to Equation (7.2). The amplitude of the $^1O_2$-luminescence kinetics at each grid point is shown in Figure 7.3 (right). The scanning procedure took about 25 min and yellow pixel in the resulting image represent higher amplitudes. This shows a nice agreement between the blood vessels in the photograph with the higher amplitudes.

For the information available from Figure 7.3, in principle one could have been using an NIR-sensitive camera to capture a steady-state intensity image. That would have drastically reduced the imaging duration to about 2–3 s and increased the spatial resolution. Nevertheless, acquiring a temporal resolution would not have been possible.

> Without time resolution it is much harder to undoubtedly proof the presence of $^1O_2$. The intensity alone can give a false impression on the efficacy of $^1O_2$ generation. Additional information about the effectiveness of $^1O_2$ generation and the photodynamic action can only be gained from the kinetics.

With the kinetics analysis, however, one can clearly distinguish between the blood vessels with high- or low-oxygen partial pressure. That big difference is shown in Figure 7.4. The decay time in the oxygen-rich blood of the veins is significantly shorter than in arteries whereas the rise time is about the same. Therefore, it could be shown that the decay time of the kinetics is associated with the triplet decay time of the PS.

A detailed analysis of the mean PS triplet decay times of 41 blood vessels is shown in Figure 7.5. More data from veins is available since they usually are closer to the surface of the CAM and therefore easier accessible. The histogram data was fitted using a Gaussian for the decay times of the veins and the arteries.

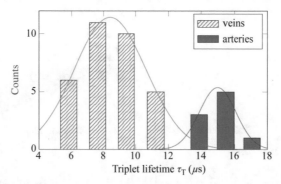

Figure 7.5: Histogram of the mean triplet lifetimes τ_T of 41 scanned blood vessels. The distributions were fitted with a Gaussian resulting in $T_{ven} = (8 \pm 3)$ µs for oxygen-rich venous blood and $T_{art} = (15 \pm 2)$ µs for oxygen-poor arterial blood [8].

From these two distributions, one can get a PS triplet decay time of $(15 + -2)$ µs for arterialized blood and $(8 + -3)$ µs for mixed venous blood. The broadening, which can be seen especially in the triplet decay times of the mixed venous blood, can be attributed to deviations between samples and especially to different oxygen saturations within the veins. Another point is the spatial overlap or close proximity of some arteries and veins. That fact, in combination with a spatial resolution of the detection system in the order of hundreds of micrometers, surely attributes to a distribution with a certain overlap between venous and arterial PS triplet decay times.

Most recent analysis of data with further increased SNR showed deficiencies in the goodness of fit for $N = 1$ in Equation (7.2). Therefore, it was proposed to include another mono exponential term, representing PS long-decaying signals from areas with very low oxygen availability. For the fit procedure parameter limits of $0.1 \leq \tau_1 \leq 1.5$ µs and $\tau_2 > 15$ µs to 20 µs should be applied. Unfortunately, this change in model also changes τ_Δ and especially τ_T to shorter values.

Another great method to visualize NIR scan data is the singlet oxygen lifetime analysis as proposed by Pfitzner et al. [22]. The NIR kinetics in each point are integrated over three areas, normalized and attributed to the RGB colors. The first area (red) usually contains the rising part of the $^1O_2$ kinetics as well as any STA like PS fluorescence, scattering and luminescence caused by the optical elements of the detection pathway. The second area (green) contains information about the $^1O_2$ luminescence intensity with a typical PS triplet decay time between 4 µs and 25 µs. The last area (blue) comprises long-decaying NIR signals like PS phosphorescence in poorly oxygenated areas (compare Figure 7.6).

Figure 7.6 shows a scan of an area with a large vein and artery crossing each other. Comparing the two images in the first row, one cannot find additional information from the stationary $^1O_2$ luminescence (Figure 7.6). The intensity from the integrated $^1O_2$-signal (without the

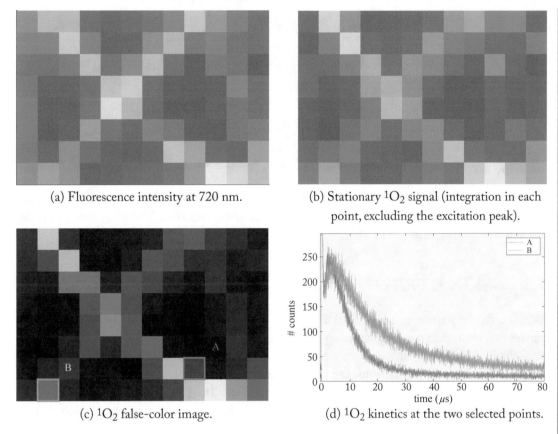

(a) Fluorescence intensity at 720 nm.

(b) Stationary $^1O_2$ signal (integration in each point, excluding the excitation peak).

(c) $^1O_2$ false-color image.

(d) $^1O_2$ kinetics at the two selected points.

Figure 7.6: Comparison between different model-free analysis methods at an exemplary crossing of an artery (low oxygenation, bottom-left–top-right) and a vein (high oxygenation, top-left–bottom-right) of the chicken CAM. Integration time for NIR data was 7 s and 200 ms for the fluorescence data.

laser peak) is roughly the same as the fluorescence intensity (at 720 nm) in the corresponding pixel. However, one has to keep in mind that the integration time for the acquisition of the NIR data is about 35 times longer than for fluorescence detection (200 ms vs. 7 s). Both images show roughly the same intensity for the oxygen-rich vein (top-left to bottom-right) and the artery (bottom-left to top-right). Therefore, the difference in oxygen saturation between arterialized and mixed venous blood cannot be visualized.

This difference in kinetics corresponding to oxygen saturation can clearly be seen in Figure 7.6 and can be nicely visualized without any data fitting, as shown in Figure 7.6. That kind of visualization shows important information about the kinetics as well as the intensities and allows for differentiation of different types of tissue.

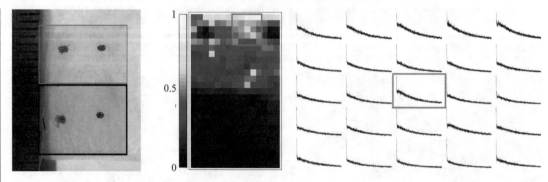

Figure 7.7: $^1O_2$ phosphorescence signals from pig ear skin (left), normalized intensity on the area indicated by the black dots (middle), and kinetics of the indicated 5 x 5 pixel area, full scales are 400 counts and 35 µs [24].

7.3   SKIN *EX VIVO*

Skin is a very important *ex vivo* model to examine uptake effects into tissue through the natural barrier that surrounds most higher living beings. First experiments with $^1O_2$ on skin were already published in 2002 [23]. This relatively early publication of results on $^1O_2$ in tissue was mainly facilitated by the following reasons: On the one hand, skin is easily accessible and absorption and scattering by tissue can be neglected during measurement. The PSs can also be applied directly to or injected into the area of interest. Finally, the $^1O_2$ signal in skin is much stronger than in many *in vivo* systems. These facts enabled 2D scans of time resolved $^1O_2$ phosphorescence on skin *ex vivo* with a resolution of up to 45 px/inch.

Schlothauer et al. developed a 2D scanning procedure and the PS was topically applied to the samples using a vanishing cream [24]. After a certain time, the skin was washed to get rid of residual cream and PS. Using a multi-furcated fiber to scan an area of 11.2 by 6.5 mm, the authors succeeded in clearly distinguishing between treated and non-treated areas. An excitation at 666 nm with 11 mW was enough to get clear kinetics at each pixel after only 10 s of integration time (Figure 7.7).

A clear correlation of the $^1O_2$ phosphorescence intensity and the topography of the skin could be found. Structures like hair follicles and furrows seem to be accumulation sites for the PS and show a higher $^1O_2$ phosphorescence intensity. Furthermore, the evaluation of the decay times indicates that in the hair follicles the kinetics clearly differ from the one in cream, indicating that the stronger signals form follicles are not only caused by an accumulation of PS cream.

The authors found surprisingly high $^1O_2$ phosphorescence decay times in skin ranging from 12.5–19.2 μs, which are at least 3.5 times longer than the decay time in water. The attribution of decay times to $^1O_2$ and PS triplet decay time was done via additional measurements in corneocytes and under nitrogen atmosphere [24, 25]. The low water content, the scarcity of chemical quenchers, and the intercellular lipids that make up 20% of the stratum corneum are all possible reasons for a prolonged $^1O_2$ decay time. However, without investigations toward the content of unsaturated lipids and the efficacy of biological quencher molecules in the skin, the exact location of $^1O_2$ generation cannot be determined.

The investigations clearly showed that the PSs in question do not penetrate healthy skin. Only after impairing the skin via tape stripping, some PSs were found deeper in the tissue [25]. This provides means of selective delivery for PDT treatment of skin conditions like psoriasis or skin cancer, which increase skin permeability due to an enhanced proliferation.

> The thickness of skin and its interaction with $^1O_2$ influence the signal strength of $^1O_2$ phosphorescence. As a result, the signal in the upper skin layers is comparatively strong and becomes much weaker when PSs penetrate deeper into the tissue and $^1O_2$ interacts with the cells. These interactions can only be visualized by time-resolved NIR measurements.

7.4   MOUSE

Rodents are used today for a variety of pharmaceutical, therapeutic and imaging examinations [26]. Since they are fully developed animals, the ethical barriers to investigation are much higher than in the CAM model. However, in contrast to the latter model, rodents can also be used to investigate systemic effects.

To date, only few research groups worldwide reported direct $^1O_2$ luminescence measurements *in vivo*; the first direct $^1O_2$ luminescence detection *in vivo* dates back to 2002. Niedre et al. reported about $^1O_2$ generated by AlS_4Pc in water, from suspensions of murine leukemia cells as well as from the liver of a rat [23]. While Niedre et al. observed the typical $^1O_2$ luminescence kinetics in water and from the cell suspension, they weren't able to detect it from inside the rat liver. The analysis of these early *in vivo* kinetics suffered from a low SNR. In addition, an inexplicable signal was observed to subside in the first $10\,\mu s$, so the authors monoexponentially fitted the signal between 10 and $60\,\mu s$, resulting in a decay time of $(30 \pm 5)\,\mu s$. Later experiments of this group showed that the best fitting results required a model with six parameters representing different microenvironments [27].

Later, in 2008, Lee et al. showed the feasibility of diode lasers as excitation source for time resolved $^1O_2$ luminescence [28]. The authors discussed the effects of very long excitation periods of around $10\,\mu s$ for homogeneous systems, resulting in a numeric model with which temporal profiles of PSs in aqueous solution were predictable. With these longer excitation periods—and therefore increased light doses—it was possible to see some signal at $1270\,nm$ for mice injected with about $1\,mg/kg$ Ce6. Nevertheless, the SNR of those *in vivo* measurements did not allow for an analysis of decay times.

Another eight years later, it was finally possible to acquire *in vivo* $^1O_2$ luminescence kinetics with sufficient SNR for a kinetics analysis [7]. Pfitzner et al. used a multi-furcated fiber in combination with 660 nm laser diodes and excitation periods of about 400 ns for simultaneous time resolved NIR and spectrally resolved fluorescence detection in NMRI nu/nu mice [7]. The authors succeeded in investigating $^1O_2$ kinetics following different types of PS administration. The strongest signals were achieved by topical application of the PSs. Kinetics with good SNR were also measured after subcutaneous or even intravenous PS injection. All measurements could be performed without opening the skin and are therefore suitable for dosimetry during PDT.

Only recently, a more detailed analysis of the $^1O_2$ kinetics in mouse sarcoma after systemic application of nanoparticulate PSs was possible. The $^1O_2$ phosphorescence could be discriminated from interfering PS phosphorescence, and the signals from the tumor site differ from those of healthy tissue. However, these measurements identify local oxygen supply as one of the major issues of PDT in solid tumors and underline the importance of $^1O_2$ based real-time monitoring during the treatment process of any PDT, which is about to become available soon.

In the following sections, we will examine these results, which are revolutionary for understanding the mechanism of action of PDT, a little closer.

SKIN WITH TOPIC APPLICATION

As already written, the topical application of PSs leads to the strongest $^1O_2$ luminescence signals and is therefore technically the easiest to achieve. In contrast to the *ex vivo* skin model, Ce6 dissolved in ethanol was topically administered to the back of mice. The $^1O_2$ kinetics were determined in successive measurements, each with an integration time of 10 s, so that the development of the decay times could be monitored over the course of the experiment. The data were adjusted with Equation (7.1), although deviations of the data from the model were detected. The resulting kinetics are very similar to those observed on skin *ex vivo* [24, 25]. With increasing illumination dose, there is an initial rise in $^1O_2$ lifetime and PS triplet decay time (see Figure 7.8) before they reach a plateau after a dose of approximately 4 J/cm^2.

Since the PS triplet decay time is strongly dependent on oxygen saturation, this increase indicates a loss of oxygen in the microenvironment of the PS due to chemical quenching of the $^1O_2$. On the other hand, the increasing $^1O_2$ decay time indicates a consumption of chemical $^1O_2$ quenchers [29]. Most major components of the skin like proteins, lipids, and endogenous antioxidants are known to react with oxygen and are therefore potential $^1O_2$ quenchers [30, 31]. In the *in vivo* skin, it appears that the $^1O_2$ decay time is somewhat shorter than obtained in the *ex vivo* skin model [24, 25]. This would be reasonable, as the healthy skin of the living mouse will likely have more oxygen scavengers than an *ex vivo* pig ear skin.

The deviation of the signal from the typical $^1O_2$ kinetics can be treated by equating the signals with Equation (7.2). This is illustrated in Figure 7.9 as an example of the beginning and end of the measurement presented in Figure 7.8. Such an analysis shows that, in addition to the typical $^1O_2$ kinetics, a rapidly decaying component can be identified.

Unfortunately, the physical meaning of the heuristically introduced monoexponential decay is not yet clear. Although the introduction of an additional monoexponential term

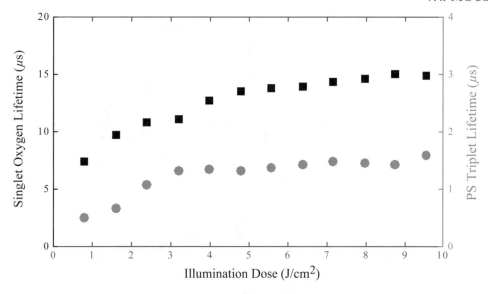

Figure 7.8: Variation of $^1O_2$ (black squares) and Ce6 (red dots) triplet decay times upon topic PS application. Fitted parameters (Equation (7.1)) range from $\tau_T = 0.5\,\mu s$ and $\tau_\Delta = 7.4\,\mu s$ to $\tau_T = 1.6\,\mu s$ and $\tau_\Delta = 14.9\,\mu s$ [7].

changes the lifetimes τ_Δ and τ_T, the already established correlation between decay time and exposure dose remains. The fit to this data with Equation (7.2) (N = 2) yields decay times of $(0.3 \pm 0.1)\,\mu s$, $(0.7 \pm 0.1)\,\mu s$, and $(9.8 \pm 0.2)\,\mu s$ at the beginning and $(0.6 \pm 0.1)\,\mu s$, $(2.0 \pm 0.1)\,\mu s$, and $(14.2 \pm 0.4)\,\mu s$ at the end of the measurement.

SUBCUTANEOUS INJECTION

Although the influence of tissue is significantly greater, clear $^1O_2$ luminescence kinetics can also be acquired after subcutaneous injection of Ce6. The typical biexponential shape can be seen in Figure 7.10, even though a small disturbance signal in the first few microseconds, which is slightly more pronounced at 1230 nm, was observed. The measurements shown here were taken two hours after subcutaneous injection of Ce6 ($0.25\,\mathrm{mg\,kg^{-1}}$ body weight) at 1270 nm and 1230 nm. The signal at 1270 nm shows typical characteristics of a $^1O_2$ phosphorescence signal (compare with inset) such as a signal rise and decay. However, the simple biexponential behavior does not suffice to describe the kinetics. At least one additional slowly decaying component has to be added for fitting the data. Doing this according to Equation (7.2) one gets a decent fit of the data with decay times of $\tau_1 = (10.9 \pm 0.3)\,\mu s$, $\tau_2 = (1.2 \pm 0.1)\,\mu s$, and $\tau_3 = (184 \pm 21)\,\mu s$ yielding the residues shown below the actual data. At 1230 nm, however, a simple biexponentially decaying signal is observed, whose decay times of $(9 \pm 1)\,\mu s$ and $(175 \pm 16)\,\mu s$ are very similar to the decay times observed at 1270 nm.

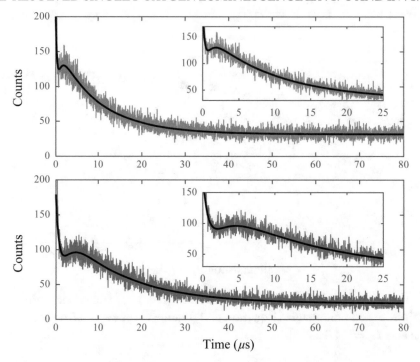

Figure 7.9: Exemplary kinetics at the begin (top) and at the end (bottom) of the measurement shown in Figure 7.8 upon topic application of Ce6. Data fitted according to Equation (7.2) yields decay times of $\tau_1 = 9.8\,\mu s$, $\tau_2 = 0.7\,\mu s$, $\tau_3 = 0.3\,\mu s$ (top) and $\tau_1 = 14.2\,\mu s$, $\tau_2 = 2.0\,\mu s$, $\tau_3 = 0.6\,\mu s$ (bottom) [7].

Based on this observation, we can assume that the 1270 nm signal is the sum of a spectrally broader disturbance signal and the $^1O_2$ phosphorescence signal. Since the slowly decaying component shown in Figure 7.10 is also detectable at 1230 nm with almost the same intensity as at 1270 nm it is very unlikely to be a pure $^1O_2$ phosphorescence, which should be significantly less intense at 1230 nm. Nevertheless, the presence of Ce6 as photosensitizer must cause it directly, because no such signal could be detected before injection. It should be noted that such long decaying signals were also observed about 1.5 h after intravenous injection of Ce6 at both detection wavelengths when measured over larger blood vessels, especially on the mouse flank (see Section 7.4).

An obvious approach for the data evaluation would be to subtract the 1230 nm signal from the one centered at 1270 nm. The difference indeed looks like a typical $^1O_2$ signal, as shown in Figure 7.11. The distribution of residuals shows that a fit with the biexponential model (Equation (7.1)) describes the data very well. The reduced χ^2 is 0.97, the decay times are the decay times are $\tau_1 = (10.4 \pm 0.3)\,\mu s$ and $\tau_2 = (1.6 \pm 0.1)\,\mu s$. These decay times differ significantly

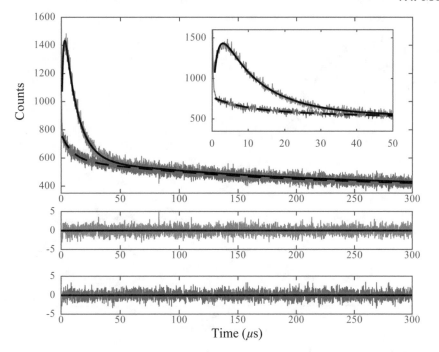

Figure 7.10: Luminescence signals measured at 1272 nm (dark gray) and 1233 nm (light gray) 2 h after subcutaneous injection of Ce6 (0.25 mg kg^{-1}). The fit at the $^1O_2$ wavelength was obtained by using the model according to Equation (7.2) yields the parameters: $\tau_1 = 10.9\,\mu s$, $\tau_2 = 1.2\,\mu s$, and $\tau_3 = 184\,\mu s$. A bi-exponential decay fit to the data at 1233 nm yields the following parameters: $\tau_1 = 9.4\,\mu s$, $\tau_2 = 175\,\mu s$ [7].

from the ones measured with Ce6 in PBS ($\tau_\Delta = 3.5\,\mu s$, $\tau_T = 2\,\mu s$), so we have to consider an interaction between the $^1O_2$ and its environment that is different from that in water. Interestingly, the determined decay times of around 10 µs and 1 µs are very similar to the ones measured for topical application of Ce6. However, this does not necessarily mean that the assignment of the fitted decay times to the physical parameters is the same.

TUMOR

Intravenous injection of PS is a common clinical practice for the treatment of solid tumors. However, concerning the detection of $^1O_2$ luminescence and data evaluation, these are probably the most difficult experimental conditions.

In the following, Ce6-based Polyamidoamine (PAMAM) dendrimers were used as PS, as their cellular uptake is significantly higher than that of free Ce6 [32]. Since the dendrimer-

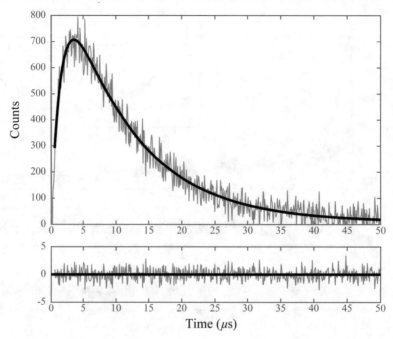

Figure 7.11: Difference of signals at 1272 nm and 1233 nm shown in Figure 7.10. Data was fitted using Equation (7.1) yielding decay times of $\tau_1 = 10.4\,\mu s$ and $\tau_2 = 1.6\,\mu s$ [7].

bound Ce6 requires a longer incubation time, measurements were conducted 16 h after IV injection of 5 mg kg^{-1} body weight Ce6-dendrimers solved in PBS.

Figure 7.12 shows the kinetics measured through the skin in the area of the tumor. The data were recorded with filters at 1270 nm and 1233 nm and an integration time of 50 s each. The fit of the data recorded at 1230 nm yields an almost monoexponential decay ($\tau_1 = (0.5 \pm 0.1)\,\mu s$) with a small second component (< 5%) amplitude and $\tau_2 = (10 \pm 3)\,\mu s$. To minimize the number of free parameters, the data recorded at 1270 nm were fitted according to Equation (7.2), resulting in decay times of $(5.8 \pm 0.3)\,\mu s$, $(1.0 \pm 0.2)\,\mu s$, and $(0.5 \pm 0.1)\,\mu s$. The first two times correspond to those of the biexponential part of the model and the third additional time shows good agreement with the dominant part of the kinetics determined at 1230 nm. Subtraction of the Signal at 1230 nm from the one at 1270 nm and subsequent fit with the biexponential model yields the same decay times for this component.

So far, it has been difficult to assign the fitted decay times to the physical properties of the sample such as the triplet or $^1O_2$ decay times. When considering possible $^1O_2$ decay times *in vivo*, it is extremely important to know the microenvironment of the PS. The first assumption is that this would be an aqueous environment, since water is present in all cells and Ce6 is a water-soluble PS. In this case, the $^1O_2$ decay time under these conditions would have an upper limit

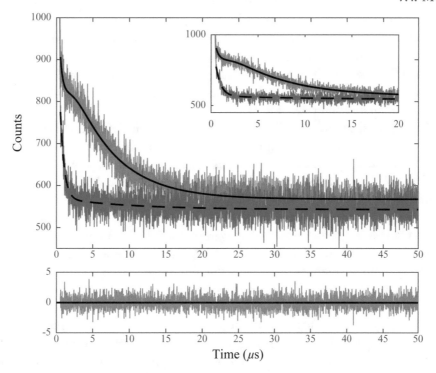

Figure 7.12: Luminescence signals at 1272 nm (red) and 1233 nm (blue) measured 16 h after IV injection of Ce6 dendrimer at the tumor area through the skin [7].

of about 4 µs. The decay times of 5.8 µs and 10.4 µs resulting from the data measured during IV injection of PS in the tumor area or during subcutaneous injection can therefore be regarded as triplet decay times. These longer decay times compared to Ce6 in water could be explained by a reduced oxygen availability in the tissue [33]. Due to the large detection spot, one can expect to detect luminescence generated in different environments at the same time. Therefore, it is not surprising to find additional components in the luminescence kinetics like the long ones described here.

In order to better distinguish $^1O_2$ phosphorescence from other interfering signals, new detection optics with high transmission and spectral position of the central wavelength at 1200, 1270, and 1340 nm were developed for this application. Although the full width half maximum (FWHM) of the filters is about 30 nm, the use of two optics in addition to the typical $^1O_2$ wavelength allows to safely subtract the thermal background of the detection system and the PS phosphorescence as long as its relative intensity is determined independently at the three spectral positions of the optics. In this way, the observed slowly decaying component can be analyzed in particular.

Under conditions with low oxygen saturations and high quencher concentrations, the PS phosphorescence and $^1O_2$ phosphorescence kinetics are difficult to distinguish. The clearest difference in kinetics is only visible shortly after the excitation pulse, i.e., exactly the time range masked by the STA (see Subsection 4.2.2). Nevertheless, one can try to estimate the PS phosphorescence intensity at 1270 nm from the measurements at 1200 nm and 1340 nm and attribute the difference to the measured signal to $^1O_2$.

The temporal shape of the STA could be isolated from the detected signal at 1340 nm by subtracting only the thermal background and the monoexponential PS phosphorescence. Although the physical significance of this signal is not yet known, a number of exponential decays are probably a good description. For the given SNR, a double exponential decay was appropriate.

The detection optics described above were used for the injection of HPMA pyropheophorbide-a conjugates into the tail vein of mice. HPMA conjugates are nanoparticulate drugs which exploit the so-called *Enhanced Permeability and Retention* (EPR) effect [34] and therefore accumulate highly selectively in the tumor region (mouse sarcoma) [35]. Despite this property, in this experiment the $^1O_2$ signal determined outside the tumor region is stronger than in the tumor region.

For the analysis of the signal from the tumor, it was necessary to integrate the already described STA into the adaptation. Background and PS phosphorescence corrected kinetics inside and outside the tumor area both comprise a $^1O_2$ signal with almost identical decay times at both sites, but a lower amplitude in the tumor Figure 7.13. In addition, both signals contain a small, very long monoexponential decay. Interestingly, the $^1O_2$ kinetics do not change with prolonged illumination. The analysis of the drug distribution after sacrifice showed high drug amounts in blood plasma, but no drug in skin or muscles. It can therefore be assumed that the determined signals originate from drugs circulating in the bloodstream.

In the tumor, high amounts of extravasated drug reduce the local penetration depth of the laser excitation, which could explain the lower amplitude of the signal from the tumor. Results of PDT on cell cultures also suggest that the oxygen consumption by PDT clearly exceeds that of the natural metabolism of tumor cells. The cell metabolism alone reduces the local oxygen partial pressure to 2% [36], the additional oxygen consumption during PDT thus exceeds any possible oxygen supply to the tissue [37]. Consequently, the missing $^1O_2$-phosphorescence of the extravasated drug can be explained by rapid oxygen depletion under illumination.

Experimental results show that the monitoring of PDT *in vivo* could become possible in the near future. As of now, it seems that the most important information obtained in this way is the local availability of oxygen. Nevertheless, it is of great advantage to distinguish active PSs from those exposed to low oxygen partial pressure in order to optimize the dose and duration of PDT treatment in the future.

Whether or not more detailed information may be determined in the future mainly depends on further improvements in the sensitivity of the detection setup and the detector itself.

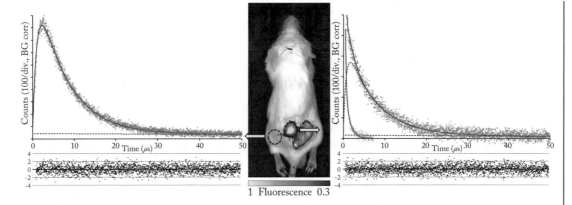

Figure 7.13: Singlet oxygen phosphorescence kinetics outside the tumor (left) and at the tumor (right) after 72 h. Duration of detection was 100 s each. The tumor data comprise some typical STA. The dashed curves represent the signal components corrected for the short time artifact. Corresponding PS fluorescence intensity is shown in the middle (kinetics reprinted from [37]).

REFERENCES

[1] T. Patrice. *Photodynamic Therapy*. Comprehensive Series in Photochemical & Photobiological Sciences. Cambridge: Royal Society of Chemistry, 2003. DOI: 10 . 1039 / 9781847551658.

[2] J. Yamamoto et al. Monitoring of singlet oxygen is useful for predicting the photodynamic effects in the treatment for experimental glioma. *Clinical Cancer Research : An Official Journal of the American Association for Cancer Research* 12.23 (2006), pp. 7132–7139. DOI: 10.1158/1078-0432.CCR-06-0786.

[3] S. Lee et al. Dual-channel imaging system for singlet oxygen and photosensitizer for PDT. *Biomedical Optics Express* 2.5 (2011), pp. 1233–1242. DOI: 10.1364/BOE.2.001233.

[4] S. Mallidi et al. Photosensitizer fluorescence and singlet oxygen luminescence as dosimetric predictors of topical 5-aminolevulinic acid photodynamic therapy nduced clinical erythema. *Journal of Biomedical Optics* 19.2 (2014), p. 028001. DOI: 10.1117/1.JBO.19.2. 028001.

[5] M. J. Niedre, M. S. Patterson, and B. C. Wilson. Direct near-infrared luminescence detection of singlet oxygen generated by photodynamic therapy in cells in vitro and tissues in vivo. *Photochemistry and Photobiology* 75.4 (2002), pp. 382–391. DOI: 10.1562/0031-8655(2002)0750382DNILDO2.0.CO2.

[6]   A. Preuß et al. Photodynamic inactivation of mold fungi spores by newly developed charged corroles. *Journal of Photochemistry and Photobiology B: Biology* (2014), pp. 39–46. DOI: 10.1016/j.jphotobiol.2014.02.013.

[7]   M. Pfitzner et al. Prospects of in vivo singlet oxygen luminescence monitoring: Kinetics at different locations on living mice. *Photodiagnosis and Photodynamic Therapy* 14 (2016), pp. 204–210. DOI: 10.1016/j.pdpdt.2016.03.002.

[8]   A. Looft et al. In vivo singlet molecular oxygen measurements: Sensitive to changes in oxygen saturation during PDT. *Photodiagnosis and Photodynamic Therapy* (2018). DOI: 10.1016/j.pdpdt.2018.07.006.

[9]   D. Ribatti. The chick embryo chorioallantoic membrane (CAM). A multifaceted experimental model. *Mechanisms of Development* 141 (2016), pp. 70–77. DOI: 10.1016/j.mod.2016.05.003.

[10]   K. Kunzi-Rapp, H. Schneckenburger, and C. Westphal-Frösch. Test system for human tumor cell sensitivity to drugs on chicken chorioallantoic membranes. *In Vitro Cellular & Developmental Biology - Animal* 28.9 (1992), pp. 565–566. DOI: 10.1007/BF02631021.

[11]   D. Ribatti. *The Chick Embryo Chorioallantoic Membrane in the Study of Angiogenesis and Metastasis*. Springer Science & Business Media, 2010.

[12]   A. S. Kishore et al. Hen egg chorioallantoic membrane bioassay: an in vitro alternative to draize eye irritation test for pesticide screening. *International Journal of Toxicology* 27.6 (2008), pp. 449–453. DOI: 10.1080/10915810802656996.

[13]   P. Nowak-Sliwinska, T. Segura, and M. L. Iruela-Arispe. The chicken chorioallantoic membrane model in biology, medicine and bioengineering. *Angiogenesis* 17.4 (2014), pp. 779–804. DOI: 10.1007/s10456-014-9440-7.

[14]   Z. Qiu et al. Monitoring blood volume fraction and oxygen saturation in port-wine stains during vascular targeted photodynamic therapy with diffuse reflectance spectroscopy: Results of a preliminary case study. *Photonics & Lasers in Medicine* 3.3 (2014), pp. 273–280. DOI: 10.1515/plm-2014-0012.

[15]   H. Qiu et al. Vascular targeted photodynamic therapy for bleeding gastrointestinal mucosal vascular lesions: a preliminary study. *Photodiagnosis and Photodynamic Therapy* 9.2 (2012), pp. 109–117. DOI: 10.1016/j.pdpdt.2011.11.003.

[16]   A. Kawczyk-Krupka et al. Vascular-targeted photodynamic therapy in the treatment of neovascular age-related macular degeneration: Clinical perspectives. *Photodiagnosis and Photodynamic Therapy* 12.2 (2015), pp. 161–175. DOI: 10.1016/j.pdpdt.2015.03.007.

[17]   A. Kawczyk-Krupka et al. Treatment of localized prostate cancer using WST-09 and WST-11 mediated vascular targeted photodynamic therapy - a review. *Photodiagnosis and Photodynamic Therapy* 12.4 (2015), pp. 567–574. DOI: 10.1016/j.pdpdt.2015.10.001.

[18]   D. Ribatti et al. Chorioallantoic membrane capillary bed: a useful target for studying angiogenesis and anti-angiogenesis in vivo. *The Anatomical Record* 264.4 (2001), pp. 317–324.

[19]   E. I. Deryugina and J. P. Quigley. Chick embryo chorioallantoic membrane model systems to study and visualize human tumor cell metastasis. *Histochemistry and Cell Biology* 130.6 (2008), pp. 1119–1130. DOI: 10.1007/s00418-008-0536-2.

[20]   A. J. Johnson et al. Comparative analysis of enzyme and pathway engineering strategies for 5FC-mediated suicide gene therapy applications. *Cancer Gene Therapy* 18.8 (2011), pp. 533–542. DOI: 10.1038/cgt.2011.6.

[21]   M. Klingenberg et al. The chick chorioallantoic membrane as an in vivo xenograft model for Burkitt lymphoma. *BMC Cancer* 14.1 (2014), p. 265. DOI: 10.1186/1471-2407-14-339.

[22]   M. Pfitzner, A. Preuß, and B. Röder. A new level of in vivo singlet molecular oxygen luminescence measurements. *Photodiagnosis and Photodynamic Therapy* 29 (2020), p. 101613. DOI: https://doi.org/10.1016/j.pdpdt.2019.101613.

[23]   M. J. Niedre et al. Measurement of singlet oxygen luminescence from AML5 cells sensitized with ALA-induced PpIX in suspension during photodynamic therapy and correlation with cell viability after treatment. *Optical Methods for Tumor Treatment and Detection: Mechanisms and Techniques in Photodynamic Therapy XI*, ed. by T. J. Dougherty. Vol. 4612. International Society for Optics and Photonics. SPIE, 2002, pp. 93–101. DOI: 10.1117/12.469336.

[24]   J. C. Schlothauer et al. Luminescence investigation of photosensitizer distribution in skin: correlation of singlet oxygen kinetics with the microarchitecture of the epidermis. *Journal of Biomedical Optics* 18.11 (2013), p. 115001. DOI: 10.1117/1.JBO.18.11.115001.

[25]   J. C. Schlothauer et al. Time-resolved singlet oxygen luminescence detection under photodynamic therapy relevant conditions: comparison of ex vivo application of two photosensitizer formulations. *Journal of Biomedical Optics* 17.11 (2012), p. 115005. DOI: 10.1117/1.JBO.17.11.115005.

[26]   Z. S. Silva et al. Animal models for photodynamic therapy (PDT). *Bioscience Reports* 35.6 (2015), pp. 1–14. DOI: 10.1042/BSR20150188.

[27]   M. T. Jarvi et al. Singlet oxygen luminescence dosimetry (SOLD) for photodynamic therapy: current status, challenges and future prospects. *Photochemistry and Photobiology* 82.5 (2006), pp. 1198–1210. DOI: 10.1562/2006-05-03-IR-891.

[28]   S. Lee et al. Pulsed diode laser-based singlet oxygen monitor for photodynamic therapy: in vivo studies of tumor-laden rats. *Journal of Biomedical Optics* 13.6 (2008), p. 64035. DOI: 10.1117/1.3042265.

[29]   S. Hackbarth et al. New insights to primary photodynamic effects–Singlet oxygen kinetics in living cells. *Journal of Photochemistry and Photobiology B: Biology* 98.3 (2010), pp. 173–179. DOI: 10.1016/j.jphotobiol.2009.11.013.

[30]   J. J. Thiele et al. Protein oxidation in human stratum corneum: susceptibility of keratins to oxidation in vitro and presence of a keratin oxidation gradient in vivo. *The Journal of Investigative Dermatology* 113.3 (1999), pp. 335–339. DOI: 10.1046/j.1523-1747.1999.00693.x.

[31]   J. J. Thiele. Oxidative targets in the stratum corneum. *Skin Pharmacology and Physiology* 14.Suppl. 1 (2001), pp. 87–91. DOI: 10.1159/000056395.

[32]   E. Bastien et al. PAMAM G4.5-Chlorin e6 dendrimeric nanoparticles for enhanced photodynamic effect. *Photochemical & Photobiological Sciences: Official Journal of the European Photochemistry Association and the European Society for Photobiology* (2015). DOI: 10.1039/C5PP00274E.

[33]   S. Oelckers. *Singulett-Sauerstoff im Modellsystem photosensibilisierte Erythrozyten-Ghost-Suspensionen: Apparative Entwicklungen und zeitaufgelöste spektroskopische Untersuchungen.* 1. Aufl. Berlin: Mensch-und-Buch-Verl., 1999.

[34]   Y. Matsumura and H. Maeda. A new concept for macromolecular therapeutics in cancer chemotherapy: mechanism of tumoritropic accumulation of proteins and the antitumor agent smancs. *Cancer research* 46.12 Part 1 (1986), pp. 6387–6392. URL: http://cancerres.aacrjournals.org/content/46/12_Part_1/6387.

[35]   J. Fang et al. N-(2-hydroxypropyl)methacrylamide polymer conjugated pyropheophorbide-a, a promising tumor-targeted theranostic probe for photodynamic therapy and imaging. *European Journal of Pharmaceutics and Biopharmaceutics : Official Journal of Arbeitsgemeinschaft Für Pharmazeutische Verfahrenstechnik e.V.* 130 (2018), pp. 165–176. DOI: 10.1016/j.ejpb.2018.06.005.

[36]   B. Muz et al. The role of hypoxia in cancer progression, angiogenesis, metastasis, and resistance to therapy. *Hypoxia (Auckland, N.Z.)* 3 (2015), pp. 83–92. DOI: 10.2147/HP.S93413.

[37]   S. Hackbarth et al. Singlet oxygen phosphorescence detection in vivo identifies PDT-induced anoxia in solid tumors. *Photochemical & Photobiological Sciences: Official Journal of the European Photochemistry Association and the European Society for Photobiology* 18.6 (2019), pp. 1304–1314. DOI: 10.1039/C8PP00570B.

Authors' Biographies

STEFFEN HACKBARTH

Steffen Hackbarth received his Ph.D. degree in experimental physics from HU Berlin in 2000. Ever since, he has been working in the field of time-resolved spectroscopy in the time range ps to ms with a special focus at the triplet processes of photosensitizers. He is head of the singlet oxygen lab and as such, recently became a faculty member of the mathematical and natural sciences faculty. Since 2011 he is leader of the advanced practical courses for the physics students at HU Berlin. His research focuses on molecular photobiophysics, especially on fundamental research in the field of photosensitization, passive drug delivery, and diffusion processes during Photodynamic Therapy. Technical developments focus on singlet oxygen luminescence detection *in vivo* and other heterogeneous environments at highest sensitivity toward real-time supervision of individualized medical tumor treatments. In 2012, he was awarded (together with his colleague, Jan Schlothauer) the innovation award of the SPIE Europe for the development of a compact time-resolved table-top singlet oxygen luminescence detection system, which since then was improved to *in vivo* capability.

MICHAEL PFITZNER

Michael Pfitzner received his M.SC in physics from HU Berlin in 2013. Since then, he was working in the field of time-resolved spectroscopy with a special focus at singlet oxygen spectroscopy. He will finish his Ph.D. this year, with a focus at pushing *in vivo* singlet oxygen measurements toward a medical application. His special interest lies within the detection of these very weak NIR signals as well as data evaluation. In 2017, he began expanding his studies toward the combination of time-resolved pointwise measurements with steady-state camera-based detection methods at the Fujian Normal University (Fuzhou, China).

JAKOB POHL

Jakob Pohl studied biophysics at the Humboldt-Universität zu Berlin and received his Master's degree in 2013. During a three-year stipendium from Deutsche Bundesstiftung Umwelt, he developed protocols for a reproducible cultivation of phototrophic biofilms under conditions of photodynamic inactivation. Those are, among others, the major topics of his Ph.D. thesis which he completed in 2019. Since his Bachelor's thesis in 2009, he has been working in the field of photodynamic therapy and photodynamic inactivation, studying the effects of photosensitization on human cancer cells and microorganisms. Since 2016 he has worked as a research associate at the Humboldt-Universität zu Berlin in the group of photobiophysics to develop protocols for qualitative analysis of PDI, photophysical characterization of antimicrobial surfaces, and the inactivation of algal biofilms.

BEATE RÖDER

Beate Röder has been professor for experimental physics in the Department of Physics, Humboldt-Universität zu (HU) Berlin since 1993. She retired in May 2018 and now works as a guest researcher at the same institution. During her tenure, she headed the photobiophysics group and supervised about 20 Ph.D. students. She received her Ph.D. degree in experimental physics from HU Berlin in 1982 and the habilitation in 1986. Her research focuses on molecular photobiophysics, especially on fundamental research in the field of photosensitization and artificial photosynthesis. Besides these topics, her research was and is focused on the development of time-resolved optical methods for detection of very low-light intensities. She has been working for more than 30 years on direct spectroscopic detection of singlet molecular oxygen, in recent years especially on time-resolved luminescence detection. In 1986, she was awarded with the Humboldt Research Price for her habilitation in the field of fundamental research on molecular mechanisms of PDT. In 1989 she was a visiting scientist at Bowling Green State University (USA) in the group of Michael Rodgers, one of the pioneers in singlet oxygen luminescence detection. In the 1990s, she also visited the Weizmann Institute in Israel. In 2009, she was awarded a Walton Professorship by Science Foundation Ireland.